Aristotle on False Reasoning

SUNY series
in
Ancient Greek Philosophy

Anthony Preus, editor

Aristotle on False Reasoning

Language and the World in the *Sophistical Refutations*

Scott G. Schreiber

State University of New York Press

Published by
State University of New York Press, Albany

Printed in the United States of America

For information, address State University of New York Press,
90 State Street, Suite 700, Albany, NY 12207

Production by Michael Haggett
Marketing by Fran Keneston

Library of Congress Cataloging-in-Publication Data

Schreiber, Scott G. (Scott Gregory), 1952–
Aristotle on false reasoning : language and the world in the Sophistical refutations / Scott G. Schreiber
p. cm. — (SUNY series in ancient Greek philosophy)
Includes bibliographical references and index.
ISBN 0-7914-5659-5 (alk. paper) — ISBN 0-7914-5660-9 (pbk. : alk. paper)
1. Aristotle. 2. Reasoning. 3. Fallacies (Logic) I. Title. II. Series.

B491.R4 S37 2003
185—dc21 2002030968

10 9 8 7 6 5 4 3 2 1

To Sophia
ἀτμὶς γάρ ἐστι τῆς τοῦ Θεοῦ δυνάμεως

Contents

PART 3: FALLACIES OUTSIDE OF LANGUAGE

List of Abbreviations

The following are used to refer to the works of Aristotle:

Cael.	*On the Heavens*
Cat.	*Categories*
de Int.	*On Interpretation*
EN	*Nicomachean Ethics*
GA	*Generation of Animals*
G.C.	*Generation and Corruption*
HA	*History of Animals*
Meta.	*Metaphysics*
PA	*Parts of Animals*
Phy.	*Physics*
Poet.	*Poetics*
Pol.	*Politics*
Pr. An.	*Prior Analytics*
Pst. An.	*Posterior Analytics*
Rhet.	*Rhetoric*
S.E.	*Sophistical Refutations*
Top.	*Topics*

Preface

My interest in Aristotle's *Sophistical Refutations* was prompted by one extraordinarily bold claim that he makes early in the treatise. He says that there are twelve ways *and only twelve* ways by which false arguments can appear to be persuasive. How could that be, I wondered. Does not the rich history of human gullibility suggest a nearly unlimited number of ways that people can be fooled into accepting poor arguments? But Aristotle rarely makes such claims lightly. So began my close analysis of this treatise that purports to argue for and illustrate exactly those twelve ways of producing false but persuasive arguments. Aristotle constructs his twelvefold classification of fallacies from the perspective of the victim of the false reasoning. The question he asks is this: What would explain why some person finds some piece of false reasoning persuasive? The victim of the sophism must hold some additional false belief, either about language or about the world, which makes the false reasoning appear cogent to him. Aristotle's twelvefold taxonomy of false arguments, then, is based upon twelve types of false belief that lend persuasiveness to bad arguments. And these false beliefs are not just about the mechanics of proper logical form. For Aristotle, logical acumen alone is not enough to safeguard one from sophistical arguments. One also must possess the right meta-logical and metaphysical beliefs, and Aristotle believes that he has uncovered the twelve false beliefs about language and the world whose correction will protect one from being taken in by false argumentation.

Aristotle's classification of fallacies and his justification of that classification in the *Sophistical Refutations* have received little systematic study in the twentieth century. Such, however, was not always the case. From the early Greek commentators, through the Latin schoolmen of the medieval period, and into the nineteenth century, there had been a steady interest in the project of creating a complete taxonomy of reasoning errors. Why did this interest wane in the twentieth century? One factor is that the so-called linguistic turn in the Anglo-American philosophic world could no longer seriously entertain Aristotle's chief taxonomical distinction between errors based on language and errors based outside of language. The efforts of these philosophers, whether proponents of ordinary or ideal language, were to resolve philosophic problems exclusively through linguistic clarification. The assumption that this could be done left little sympathy for Aristotle's claim that certain kinds of false reasoning, themselves productive of philosophical perplexities needing resolution, could only be resolved through metaphysical clarification.

This book returns, with considerable sympathy, to Aristotle's project. My goal is to make clear the philosophical justification that Aristotle presents for his classification of fallacies. To do this, however, it is necessary to explore in some detail the numerous examples of fallacies that Aristotle uses for illustration. As happens so often in Aristotle, his examples can both clarify and confuse. Much of this book involves a close analysis of these often-elliptical illustrations of false reasoning. I recognize that there is a danger in treating so closely all of these examples. The reader might begin to lose sight of Aristotle's big picture: his justification of the overall taxonomy. If one does occasionally find oneself losing sight of the forest for the trees, I hope that the trees themselves are sufficiently intriguing, providing peripheral insights into other areas of logical theory and wider Aristotelian thought.

This need to concentrate on Aristotle's examples explains two particular features of the study: the extensive Greek citations and the sparing use of non-English secondary sources. I have tried to keep the book as accessible as possible to the "Greekless" reader. Much of what Aristotle says is very important to readers interested primarily in the history of logic or in the growing modern literature on informal fallacies. Accordingly, I have used my own translations of all the Greek references. Nevertheless, I also have included (most often in the notes) extensive citations of Aristotle's Greek. I owe this to those Greek readers of the book, because so many of Aristotle's fallacies are heavily dependent upon features of the Greek language. A further result of this dependency is that any translation of Aristotle's examples from Greek into another language can have significant consequences of either clarifying or obfuscating the fallacy being exemplified. Moreover, different modern languages will produce different transformations. What happens to Aristotle's examples when they are rendered into German or French adds a further layer of difficulty for the English reader trying to grasp Aristotle's theory. As a consequence, I have restricted my secondary sources predominantly to those written in English (the exceptions being the premodern Greek and Latin commentators). I would be remiss, however, not to mention an important addition to the modern scholarship on Aristotle's *Sophistical Refutations* that appeared late in 1995, after much of my own research had been completed. Louis-Andre Dorion has published an extensive French translation of and commentary on the entire treatise as a volume in the J. Vrin series, *Histoire des Doctrines de l'Antiquité Classique*. While my interpretations of Aristotle's examples sometimes differ from Dorion's, readers interested in a line-by-line commentary will find his study an important resource.

Introduction

Reasoning and the *Sophistical Refutations*

Aristotle on the Kinds of Reasoning

Central to Aristotle's philosophic method is his analysis of reasoning or the syllogism (συλλογισμός).[1] He defines a syllogism as "an argument in which, when certain things are set down, something different from the things set down follows necessarily by means of the things set down."[2] In *Topics* I, 1, Aristotle makes some preliminary distinctions among syllogisms. He divides them into four types, differentiated by the character of the "things set down," that is, by the character of the premises. Demonstrative reasoning (ἀπόδειξις) proceeds from true and primary premises, appropriate to the particular science, or else from theorems already derived from such true and primary premises.[3] Dialectical reasoning (διαλεκτικός) proceeds from common beliefs (ἔνδοξα), that is, premises believed by everyone or most people or by certain wise people.[4] The third kind of reasoning is false reasoning, or "eristic" (ἐριστικός). The general mark of eristic is reasoning that *appears* to be what it is not. Eristic falsely simulates other kinds of reasoning. Since the other kinds of reasoning have been distinguished by the nature of their premises, Aristotle initially defines eristic as reasoning from premises that are only *apparently* endoxic but not really so. This would seem to restrict eristic to apparent dialectical reasoning. Finally, there is false reasoning that simulates demonstrative syllogisms. These paralogisms (παραλογισμοί) are related to particular sciences but originate from false scientific premises.[5]

The clearest way, then, to understand *Topics* I, 1, is as a fourfold classification of syllogisms based entirely on the nature of the premises:

1. demonstrative reasoning from scientific premises,
2. dialectical reasoning from endoxic premises,
3. false reasoning (paralogisms) from premises only apparently scientific; and
4. eristic reasoning from premises only apparently endoxic.

As neat as this arrangement looks in *Topics* I, 1, it is not Aristotle's final word on the kinds of reasoning. He proceeds to disrupt the scheme in two ways. First, he distinguishes another type of reasoning called "peirastic" (πειραστικός), or examinational reasoning. Peirastic proceeds from some belief of the person being examined. This sort of premise differs from a dialectical premise in that (1) it must be believed by the person being examined (whereas in dialectic, an endoxon may be posited for examination, which neither participant is committed to) and (2) it need not be an endoxon (i.e., it may be an entirely idiosyncratic belief).[6] Peirastic is the closest successor to that Socratic questioning that characterized the early Platonic dialogues: an examination of someone's claim to know something. Second and more important, even in *Topics* I, 1, Aristotle wants to consider eristic as, more broadly, false or apparent *reasoning*, not just reasoning from false or apparent *premises*, whether endoxic or scientific. And so Aristotle finally settles on a disjunctive definition of eristic, as *either* reasoning from only apparent endoxa *or* apparent reasoning, whether from real or apparent endoxa.[7] This same definition is found in the *S.E.* introduction to eristic: "reasonings from apparent but not real endoxa, or apparent reasonings."[8]

For Aristotle, the mark of eristic is *appearance*. Eristic arguments simulate but fail to be real arguments. This characteristic of simulation also is one that Aristotle applies to sophists and sophistry. For example, the sophist trades on people's inability to distinguish the true from the false, the real from the merely apparent. He makes his living from his apparent wisdom rather than any real wisdom.[9] Naturally, then, the source of the sophist's success is his expertise in eristic. But Aristotle's sophist is more than a master at apparent-but-not-real argumentation. He also can produce real (i.e., valid) arguments that appear to be, but are not, relevant to the issue at hand.[10] And so there are three sources of sophistical appearances in argumentation: premises that appear to be what they are not, arguments that appear to be valid when they are not, and valid arguments that appear to be relevant to the matter at hand when they are not. Using these three appearances, separately or in combination, the sophist derives his dangerous power to deceive. But these same false appearances can arise even apart from the intent of a sophist to deceive. One of the reasons for studying sophistical arguments, says Aristotle, is that it better prepares the philosopher for conducting his own private researches; for someone who can be deceived by another person will be all the more easily deceived by the same sorts of appearances when they arise in his own thinking.[11] How, then, does one learn to recognize these false appearances, whether they are intended by another or accidentally arise in one's own study? Aristotle devotes his treatise, *Sophistical Refutations,* to answering that question.

THE *SOPHISTICAL REFUTATIONS*

Although the work *Sophistical Refutations (S.E.)* is sufficiently self-contained to be labeled a "treatise," Aristotle seems to have intended it as the closing book to the *Topics*. So, for instance, in the *Prior Analytics* (65b16), he cites *S.E.* 167b21-36 under the title of the *Topics*. And the last chapter of *S.E.* is intended as a conclusion to the whole of his treatments of both dialectic and eristic. Nevertheless, the discussions of dialectic and eristic are clearly distinct and so marked both in the beginning of the *Topics* (100a25-101a4) and in the introduction to *S.E.* (164a20-22). In the later passage, Aristotle goes on to say that *elsewhere* he has discussed didactic, dialectical, and peirastic argumentation, and that now he must begin his treatment of eristic (*S.E.* 165a38–165b11).[12]

Aristotle has two projects in *S.E.* The first is to identify the various *sources* of false reasoning. The second is to provide the reasoner who encounters false reasoning the means to *resolve* the resultant confusion engendered by the apparent but false argument. According to Aristotle, people fall victim to false reasoning, whether in the course of a dialectical exchange with another reasoner or in the privacy of their own reflections, from two general sources. False arguments are either *due to language* (παρὰ τὴν λέξιν) or *outside of language* (ἔξω τῆς λέξεως). He further specifies six distinct linguistic sources and six distinct extralinguistic sources. The diagram on the following page shows Aristotle's entire classification.[13]

In *S.E.*4-11, Aristotle describes and illustrates each type of false reasoning, repeatedly affirming the inviolable distinction between the linguistic and the extralinguistic sources of error. Commentators have not always received this distinction kindly. Often the view has been that Aristotle's division is arbitrary. Many of the examples he cites to illustrate the different species under these two principal headings seem to be just as easily categorized under a different species from the other heading. One especially strong tendency has been to see arguments outside of language as reducible to arguments due to language.[14] One goal of this book is to show *why* Aristotle refuses to allow such a reduction. His nonreductionist position is based upon his notion of a resolution. Aristotle develops that notion in the second half of *S.E.*

In *S.E.* 16, Aristotle introduces his second concern: the problem of false reasoning from the standpoint of the potential victim of the sophism rather than from the standpoint of the perpetrator. His concern is with resolutions (λύσεις) of sophistical arguments. The organization of his material on resolutions parallels his earlier format. He devotes chapters to each of the types of fallacies, both linguistic and extralinguistic, and he shows via examples and commentary how each type is to be resolved. Aristotle requires for a resolution of a false argument two things. The resolution must explain why the false

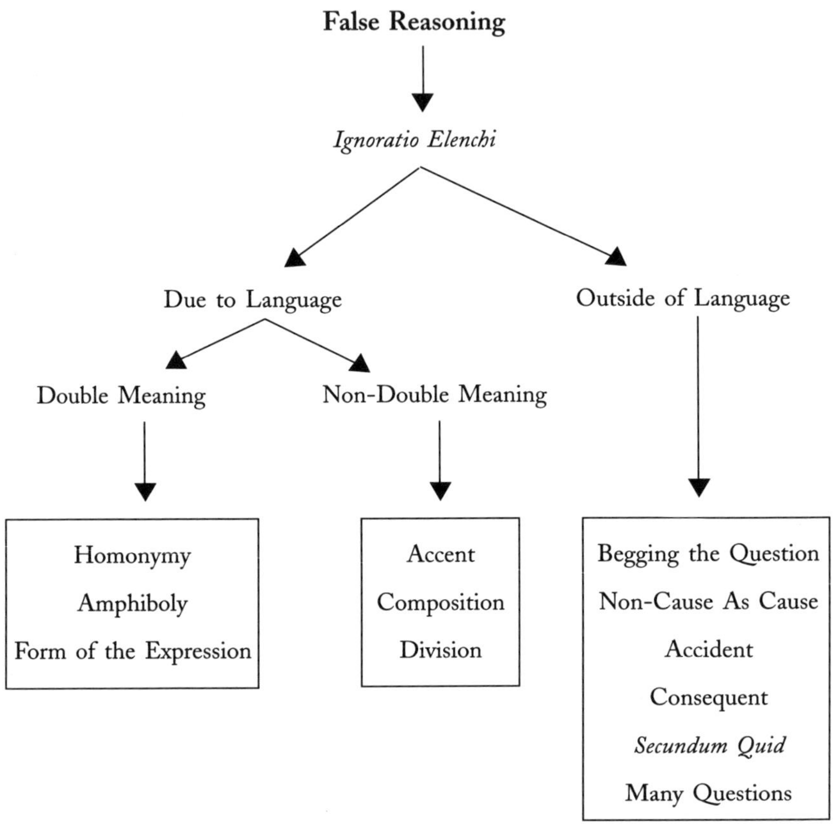

argument is false, and it must explain why it appeared to be true. It is this second explanation that plays a defining role in Aristotle's typology of fallacies. Each example of false reasoning is persuasive only if the victim holds a particular false presupposition about either language or the world. It is the nature of that presupposition that determines where the example of false reasoning is situated in Aristotle's typology.

OUTLINE OF THE BOOK

Aristotle's notion of a resolution goes a long way toward understanding his distinction between linguistic and extralinguistic fallacies. There are, however, other problems with his typology that the manner of resolution alone does not solve. Particularly on the linguistic side of the basic dichotomy, some of

Aristotle's examples raise their own peculiar difficulties. Accordingly, before considering the role of resolutions in clarifying the distinction between linguistic and extralinguistic fallacies, I analyze in part 1 Aristotle's discussions and illustrations of linguistically based fallacies. In chapter 1 I look at Aristotle's argument from *S.E.* 1, that there is a "power of names" to have multiple signification. "Multiple signification," however, turns out itself to have two meanings that Aristotle fails to keep separate. On the one hand, universals signify many different individuals as well as the universal under which the individuals fall. This is the sense of multiple signification that Aristotle shows in *S.E.* 1 to be unavoidable, given the nature and function of language. On the other hand, some words signify different *kinds* of individuals rather than just different individuals of the same kind. Both types of multivocity play roles in the production of false reasoning.

In chapters 2 and 3 I analyze the first three types of fallacy "due to language." These are the three cases of what Aristotle calls "double meaning": fallacies due to homonymy, amphiboly, and the Form of the Expression. I expose several problematic cases among Aristotle's examples of these three types. The chief source of the problems, I conclude, is Aristotle's failure to distinguish between the power of common nouns, on the one hand, both to signify universals and to apply to many particulars (as discussed in chapter 1) and, on the other hand, other kinds of multiple signification that he divides among the three fallacy types. The ways he differentiates among homonymy, amphiboly, and Form of the Expression are generally well defined and illustrated, until he tries to assign places among them to false reasonings based upon that special power of common predicates. The result is that cases of the multivocity of universal predicates end up being assigned to the various double meaning fallacy types almost arbitrarily, thereby confusing the otherwise clearly principled taxonomy. In the end I conclude that Aristotle, who fully appreciates the multivocity of so many words, fails to see (at least in *S.E.*) the multivocity of "multivocity." In my concluding chapter I will propose a revision to Aristotle's taxonomy that acknowledges the different kinds of verbal multivocity.

In chapter 4 I analyze the three fallacy types "due to language" that are not cases of double meaning: Composition, Division, and Accent. I argue that these are fallacies primarily occurring in (fourth-century B.C.) written Greek, where the absence of internal sentence punctuation, accents, breathing marks, and word divisions made it difficult for the reader to individuate separate linguistic signifiers. The same sequence of component linguistic parts (e.g., phonemes, letters, words, etc.) may turn out to compose different linguistic signifiers if enunciated differently. Errors due to Composition, Division, and Accent arise when these different signifiers are mistakenly believed to be the same signifier.

Part 2 is devoted to a general discussion of resolutions of fallacies. This section serves as the axis around which the entire book rotates, for it is the manner of resolution that determines the type of fallacy. Resolutions require the identification of those false presuppositions whose correction is both necessary and sufficient for the removal of the perplexity as to why the apparent refutation is false and why it appears true. I conclude that Aristotle recognizes three kinds of erroneous presupposition whose correction is able to resolve all perplexities arising from false reasoning. These are false beliefs about parts of language itself, false beliefs about the relationship language has to the realities it signifies, and false beliefs about the extralinguistic world that is signified. The characteristic of fallacies due to language is that their resolutions require some correction of false presuppositions about the nature of language or how language relates to the things it signifies. Resolutions of fallacies outside of language, on the other hand, require no such corrections. This is not to say, however, that the correction of errors about the nature and use of language is sufficient to resolve linguistically based fallacies. Fallacies of double meaning also derive their plausibility from particular false presuppositions about the world.

Part 3 is an analysis of the six fallacy types that arise outside of language. For each type I isolate that feature of the extralinguistic world that one must understand if one is to avoid that fallacy. In chapter 6 I argue that, for Aristotle, false reasonings due to Begging the Question and Non-Cause As Cause derive their plausibility from mistaken beliefs about the proper explanatory powers of nonlinguistic facts. In chapter 7 I discuss Aristotle's fallacy types of Accident and Consequent. I argue that Aristotle presents no convincing argument or evidence for a distinction between the two types. The common ontological mistake that renders examples of such fallacies apparently sound is the confusion of accidental with essential predication. Chapter 8 deals with the fallacy of *Secundum Quid*. I argue that these fallacies can only be resolved by correcting both false linguistic and false ontological presuppositions. Here is the most glaring taxonomic mistake in Aristotle's scheme. The need for some linguistic clarification should place these errors under Aristotle's heading of fallacies "due to language." In chapter 9 I isolate two extralinguistic errors promoting fallacies due to Many Questions. Sometimes there is a false assumption that what is truly predicable of an ontological whole or set also is truly predicable of each part of the whole or member of the set. Even where this error is not in evidence, there remains a failure to distinguish between states of affairs that are properly explanatory of some conclusion and states of affairs that only logically entail that same conclusion. In this chapter I also show that Aristotle concedes that linguistic fallacies of double meaning presuppose the extralinguistic fallacy of Many Questions. This leads to the conclusion that only the errors assigned to Composition,

Division, and Accent arise entirely independent of some mistaken ontological presupposition.

Most of the ancient and modern criticisms of Aristotle's typology of false reasonings suffer from a failure to appreciate the role of resolutions in the construction of the overall taxonomy. What emerges by the end of the book is an Aristotle whose systematic analysis of the types of false reasoning is, despite a couple of unresolved problems, principled and nonarbitrary. It rests upon a view of the world as intelligibly accessible to human understanding through the medium of (Greek) language as it is. This is not to say that language "as it is" (i.e., ordinary Greek language) is not, in both syntax and semantics, full of deceptive pitfalls for the reasoning agent. But Aristotle directs his efforts toward acquainting the human inquirer with ways to recognize those potential dangers rather than toward constructing an amended language immune to such dangers. Among Aristotle's requirements for recognizing false argumentation are commitments to a number of ontological positions. Logic, as a general study of reasoning, is not metaphysically neutral for Aristotle. He holds that there are substantive claims about the world that must be accepted if one is to be able to distinguish between examples of true and false reasoning.

Part 1

Fallacies Due to Language

Homonymy
Amphiboly
Form of the Expression
Composition
Division
Accent

Chapter 1

The Power of Names

One of the primary sources of sophistical reasoning is the equivocation between different significations of the same word or phrase within an argument. Aristotle believes that no language can avoid words of multiple signification and, therefore, that possible sophistical reasonings will be endemic to any language use. In this chapter I will show that Aristotle argues at the beginning of *S.E.* for one kind of verbal multivocity that is endemic to any language, namely, the existence of universal terms that signify both the universal and the multiple particulars under that universal. This necessary feature of language, however, is not the source of those sophistical arguments that Aristotle dwells on later in his treatise. In subsequent chapters, Aristotle will attribute most sophistical reasonings to those terms that signify different kinds of things (i.e., different universals). This kind of multivocity is not endemic to any language. In short, Aristotle conflates two sorts of verbal multivocity, one which is endemic to all language but is only rarely a cause of false reasoning, and the other which is a contingent feature of any language and is the more usual cause of false reasonings.

Naming Is Not Like Counting

In *S.E.* 1, Aristotle repeats the definition of reasoning (συλλογισμός) from *Topics* I, 1, and defines a refutation (ἔλεγχος) as reasoning to the denial of a conclusion. Attempted refutations often took place in formal dialogue between two people, referred to, in Aristotle's day, as the questioner and the answerer. The questioner was the person attempting to refute the answerer.

The questioner would begin by asking his opponent if he accepts the truth of some claim. When the answerer answered "yes," that became the proposition the questioner tried to refute. He would continue to ask the answerer if he accepts certain other claims, hoping eventually to show that these other claims agreed to by the answerer logically entailed the opposite of the original proposition. That constituted a refutation. A sophistical refutation is a line of questioning that appears to result in a refutation but is actually a fallacy (παραλογισμός) and not a refutation.[1] How do sophists produce these appearances? Aristotle says that there are many ways, but the most natural (εὐφυέστατος) and most common (δημοσιώτατος) way is through names.

> For since it is not possible to converse by bringing in the actual things themselves, but we use the names in place of the things as symbols, we think that what happens with the names also happens with the things, just as in the cases of people who calculate with counters. But it is not similar, for names and the number of expressions are limited (πεπέρανται) while the things are unlimited (ἄπειρα) in number. It is necessary then that the same expression and one name signify many things. So just as in the former case those who are not clever at handling their counters are led astray by the experts, in the same way too in the case of arguments, those who are inexperienced with the power of names miscalculate (παραλογίζονται) both in their own conversations and while listening to conversations of others.[2]

This disanalogy drawn between arithmetical counters and names (and expressions) is important, for upon it Aristotle argues for the unavoidable multivocity of language. Yet there are problems in interpreting what Aristotle means by contrasting limited names with unlimited things. If I understand correctly the force of Aristotle's claim, his disanalogy shows only the linguistic necessity of universal predicates applying to more than one individual. It does not show any necessity for predicates applying to more than one different kind of individual. To use the vocabulary of *Categories* 1, Aristotle's contrast between names and things in *S.E.* 1 only shows the necessity of synonymy, not the necessity of homonymy. To make this clearer, I must examine the purported disanalogy in some detail.

Aristotle's claim is that names are not related to the things named as counters are related to the things counted, because names are limited but things are unlimited in number. The following three questions must be addressed:

1. In what sense are names and expressions "limited"?
2. In what sense are things "unlimited in number"?

3. What does Aristotle mean by "counters," and how does the relationship between counters and what they stand for differ from the relationship between names and the things names signify?

I argue below that, for Aristotle, the number of names is limited by the number of universals, which are the proper referents for names. The names of these universals, however, possess the power to signify an unlimited number of individuals. Therefore, that "power of names," the recognition of which is so important for avoidance of fallacy, is the use of a name both to signify a universal and to apply to the particulars under that universal. I shall begin, however, with the third question and show that the relationship between counters and things counted is necessarily isomorphic in a way that the "power of names" makes impossible for the relationship between names and things named.

"Counters"

The error in assimilating names to counters, according to Aristotle, is to think that in arithmetic, as counters are to the things enumerated, so in speech, names (and expressions) are to the things signified. Those who fail to see the difference are liable to be cheated in conversation analogously to the way poor arithmeticians are cheated in calculations of prices. In short, to be fooled by an apparent analogy is to be made vulnerable to some truly analogous consequences of that false analogy! The entire example, then, provides a particularly apt introduction to the general danger of mistaking appearances for the realities that they mimic. Aristotle is warning against assimilating the activity of signifying items in the world by words or phrases to the activity of counting items in the world by counters (ψῆφοι). When Aristotle refers in the analogy to "people who calculate with counters," he probably has in mind the counters on an abacus. Arithmetical operations on an abacus were designated as "calculations by counters" (ψήφοις λογίζεσθαι). It can easily be appreciated how an inexperienced abacus user could be cheated by an unscrupulous expert. The principal point of the disanalogy with names, however, is that names are multivocal in a way that counters are not. But here one may raise an objection. Characteristic of an abacus is that the same counter can signify a different amount in different calculations. This "multivocity" of the counters on an abacus gave rise to a common Greek simile.

> [Solon] used to say that the men who surrounded tyrants were like the counters used in calculations (ταῖς ψήφοις ταῖς ἐπὶ τῶν λογισμῶν); for just as each counter signified now more and now

> less, so the tyrants would treat each of their courtiers now as great and famous, now as of no account.[3]

It is true that within each separate calculation the counter could only refer to one amount. This could provide Aristotle his contrast with names, which sophists might use to signify different things even within the same argument. There is, however, a better way to differentiate between this proverbial feature of multiple signification of counters as units in an abacus and the multiple signification of names. Even though a counter on an abacus might stand now for one unit or number, and now for another, it always stands for a definite number. In computing manpower, for instance, a counter may stand for one man, twenty men, or 100 men. It can never stand for all men or an indefinite number of men! But a name like ἄνθρωπος may refer to a particular man, or it may stand as a universal predicate, thereby signifying an indefinite number of men. The danger lurking behind the comparison, then, is to think that names, like counters, only signify particulars, either individually or in sets of limited numbers.[4]

In the mistaken analogy, counters are to the things counted as names are to the things signified. The second member of each relationship constitutes the same class. It is the class of things in the world that can be counted or signified. In both cases, they are unlimited (ἄπειρα). This cannot be understood as a claim for an actually infinite number of things, which Aristotle denies.[5] It is an appeal to the indefinite number, and thereby unknowability, of individuals that Aristotle often contrasts to the limited number of universals that are proper objects of scientific understanding.[6] The disanalogy at work between names and counters is a form of that between universals and individuals with respect to their knowability. Whereas counters are equinumerous with countable things (whether as individuals or sets of limited individuals), names and expressions are not. In the act of signifying, the absence of the isomorphism that makes computation possible is precisely what makes linguistic deception possible.

"SIGNIFIERS"

Aristotle defines "name" as "a spoken sound signifying by convention, without time, no part of which signifies in separation."[7] He includes both general terms, such as "pirate-ship" (ἐπακτροκέλης), and proper names, such as Κάλλιππος, as "names." These latter names will require some special comment below. What places limits on the number of different names in a language is the requirement that names signify (σημαίνειν). That is, the number of signifiers is restricted by the possible kinds of things that can be signified.

In his study of this relationship, Irwin argues that real, extralinguistic properties with discoverable essences are the exclusive primary objects of signification.[8] The most difficult counterexample to this position is Aristotle's claim that the nonreferring term "goat-stag" (τραγέλαφος) signifies something.[9] Irwin accounts for this by distinguishing between "signifying by nature" and "signifying to us." Although "goat-stag" fails to signify by nature, it has significance to us, that is, it signifies our beliefs about goat-stags, including the belief that no such real natures exist. By Irwin's interpretation, names that *only* signify to us have meaning without reference.

Irwin's distinction is, I think, a useful one. But his positing of a class of names that signify to us but not by nature does pose a difficulty for Aristotle's claim that names are limited. For even if names that signify by nature are limited by the limited number of real natures, the *meanings* that we can attach to nonreferring names seem to be inexhaustible. I return to this problem below. For the moment, however, let us consider how the number of names that signify by nature must be limited. Aristotle insists upon the unitary nature of any object properly signifiable by a name. According to Irwin, such a requirement explains why Aristotle denies the full status of being a name to such labels as "not-man" and "not-recovering."[10] There is no single nature common to the things that are not men or the activities that are not recovering, therefore, there is no name for such a class, only what Aristotle agrees to call "an indefinite name" (ὄνομα ἀόριστον). Names and "indefinite names" are alike in that they both signify and can be applied to multiple individuals. They differ in the presence or absence of a unitary nature common to those multiple individuals.

We know that Aristotle has restricted the number of highest kinds of things that are nameable. These are the Categories.

> Of things said without any combination, each signifies either substance or quantity or quality or a relative or where or when or being-in-a-position or having or doing or being-affected.[11]

Each name signifies by nature only one unified entity, and each such entity in turn falls into one of the *kinds* of things specified in the list of Categories. But if the number of names is truly limited by the number of nameable entities, then there also must be a limited number of *infimae species* under the higher Categories.

To illustrate how names must signify one and only one nature, Aristotle conducts a thought experiment in *de Interpretatione* 8 (18a18-27) by supposing a single term (ἱμάτιον) being given to two entities lacking a natural unity (e.g., a man and a horse). This new term is not a name, for if it signifies anything at all, then it signifies two things (a man and a horse), in much the same way that indefinite names are not strictly names because the things they signify lack a natural unity. The vexing questions of what constitute Aristotelian natural unities and how they are discovered happily need not be resolved here. It is enough to show that Aristotle believed in (1) a limited number of natural unities, and (2) that to be a name in the strictest sense was to signify one of those unities. We can now understand why the mere logical possibility of infinitely many syntactical strings recursively generable in a language would be untroubling to Aristotle when he claims that names are limited. Given any two names "A" and "B," one cannot always produce a new name (i.e., signifying by nature) "AB," since there may not exist any possible unified entity possessing such a combined nature.[12]

There remain two final obstacles to understanding Aristotle's claim that names are limited. The first deals with names of individuals and the second with names that only signify to us. Although there may be only a limited number of kinds of entities for names to signify, Aristotle also includes individuals among the entities able to be named (e.g., Κάλλιππος, *de Int.* 2, 16a21). If I am correct to interpret the contrast between things that are unlimited and names that are limited as the contrast between the unknowableness of particulars and the knowableness of universals, then the application of names to individuals seems to destroy the contrast. The same can be said about names that signify to us but not by nature. It would be possible for "goat-stag," or any nonreferring term, to signify to us. Nor would there seem to be any limit to the possible number of such names. These difficulties Aristotle never addresses. It would not be unreasonable to suppose, however, that he would regard names of individuals and names that fail to signify by nature as names in only a secondary or derivative sense. It already has been noted that so-called "indefinite names," while able to signify, are excluded from the list of names proper.[13]

This belief that only universals (i.e., essences or properties) are proper referents of names is no Aristotelian novelty. It continues a Platonic legacy wherein the primary referents of names were the Forms. Only by secondary applications were sensible particulars given the same names as the Forms they share in.[14] This, and the fact that names were regarded as somehow *naturally* connected to their universal referents, meant that, for Plato, only the philosopher or true dialectician could properly apply language to sensibles, for only he had knowledge of the Forms.[15]

In matters of linguistic derivation, Aristotle remains true to the Platonic position that names are most properly signifiers of universals.[16] In matters of ontological dependence, however, Aristotle has reversed Plato's priorities. As a result, although names primarily signify universals, and particulars are only named derivatively, those universals themselves are ontologically dependent upon those particulars. For Aristotle, then, it is the opposite directions of priority between the activity of naming and that of being that help set up the *S.E.* 1 disanalogy. The limited number of names reflects the linguistic priority of their application to universals, while the unlimited number of things reflects the ontological priority of individuals to universals. Given, then, the unlimited number and unknowable nature of individuals, names possessing the power of multiple signification become necessary epistemological tools for understanding. But this sort of multiple signification is nothing more than the power of common predicates to signify multiple individuals. It does not require that common predicates signify multiple *kinds* of individuals. This latter phenomenon, however, turns out to be one of the chief culprits among Aristotle's examples of fallacies based on linguistic double meanings.

The power of multiple signification includes for Aristotle both (nonhomonymous) universals that apply to multiple individuals of the same definition[17] and homonymous names that signify things having different definitions. The former is a necessary feature of language based on the nonisomorphic relationship between names and things signifiable, while the latter is a purely contingent feature of any given language. Yet Aristotle sometimes conflates the two. In both types of false reasoning, those generated by universals having references to multiple individuals and those generated by universals signifying different kinds of individuals, there is a failure to signify the same thing (whether individual or kind) by the same word or phrase, and this seems to have been what impressed Aristotle more than the difference between the two. This running together of these two types of multivocity explains, for instance, the strange remark in *Generation and Corruption* I, 6, which introduces his discussion of contact:

> Just as almost every other name is said in many ways, some homonymously and others from different and prior senses, so it is with "contact."[18]

It is certainly not Aristotle's claim that almost every name is "said in many ways" by being either homonymous or related to some prior focal meaning. What is true is that almost every name is "said in many ways" by applying to many particulars. That is the only sense of multiple signification that could be claimed for "almost all names."[19] The use of σχεδὸν may be Aristotle's way of qualifying the claim in recognition of exceptions such as the derivative names of individuals, or universal names such as "sun," which only happen to apply to one individual.[20]

Ultimately, if language is to be a means of human understanding of the world, the only necessary type of multiple signification of words is that of universals applying to many individuals having the same definition. Without that power, much of reality would remain hidden from the discursive probing of man. And because man naturally desires to understand, and understanding is discursive, such a state of affairs would render the universe a place of ultimate frustration for the human thinker. It is in this sense that words must possess the power of multiple signification if the universe is to be brought under the linguistic control of the human thinker. When this power of names is either intentionally abused by the sophist or just misunderstood by the inexperienced speaker, the attaining of man's final good as an understander is threatened. So it is a task of paramount importance for Aristotle to expose the misuse of this power and to explore the proper use of it.

CONCLUSION

False reasoning is persuasive insofar as it simulates true reasoning. Sophists are particularly adept in making false reasoning look true. One tactic of the simulation is to take advantage of a particular feature of language, a power of names for multiple signification. But "multiple signification" itself signifies different phenomena for Aristotle. In *S.E.* 1, he argues that in one sense multiple signification is a necessary feature of language. The basis of this necessity is the nonisomorphic relationship between limited names and unlimited things. This particular power of multiple signification is not a deficiency of language; without it, language would fail to meet the human need to attain knowledge of his world. This power, which is necessary for human understanding but holds the potential for misunderstanding through deceptive reasoning, is the power of the same common predicates both to signify a universal and to apply to separate individuals. Aristotle's argument in *S.E.* 1, however, supported by the analogy drawn between names and counters, does not entail the necessity of either homonymy or πρὸς ἕν multivocity. There is no need for the same names applying to different kinds of things, only for the same names applying to many different things of the same kind. What I show in the following two chapters is that Aristotle conflates the power of names necessary for understanding and other bases of linguistic multivocity, classified as types of "double meaning."

Chapter 2

Homonymy and Amphiboly

Introduction: Aristotle's Use of λέξις

Throughout this book I translate Aristotle's word λέξις by "language." The generality of such a rendering I consider a virtue, for it is my task here to uncover Aristotle's precise sense of the λέξις/non–λέξις dichotomy as it relates to sources of false reasoning. In the hands of later Greek writers on rhetoric and grammar, the term becomes increasingly narrowed to various technical specifications. Although Aristotle is one of the movers in that direction, it would be premature in this book (and historically anachronistic) to render his use of the word by one of the narrower terms of art that crystallized only after his death.[1] It is relevant, however, to consider the general use of the term by his philosophical mentor.

Plato uses λέξις to refer to speech in several contexts. Sometimes it is contrasted to action (πρᾶξις);[2] and sometimes it is contrasted to song (ᾠδή).[3] More narrowly, it is used to refer to a particular *style* of speech, such as that appropriate to law courts[4] or that used by poets.[5] It is this latter sense of a style or way of speaking that dominates Aristotle's use of the word in the *Poetics* and *Rhetoric*. Aristotle, like Plato, uses λέξις chiefly for oral speech, not for writing. This distinction gradually fades as the written word gains importance within the oral culture of Greece. I argue below that we find in Aristotle's fallacies of Composition, Division, and Accent reflections of just such a shift from language as an oral phenomenon to language as a written phenomenon. As a rule, however, Aristotle still considers oral speech the proper domain of λέξις. Because the English word "language" combines the same dominant sense of speech with the secondary sense of writing, I prefer it as a rendering of Aristotle's λέξις.

THE SIX SOURCES OF FALSE REASONING DUE TO LANGUAGE (παρὰ τὴν λέξιν)

There are, according to Aristotle, exactly six ways of producing the illusion of argument with language. They are:

1. Homonymy (ὁμωνυμία)
2. Amphiboly (ἀμφιβολία)
3. Composition (σύνθεσις)
4. Division (διαίρεσις)
5. Accent (προσῳδία)
6. Form of the Expression (σχῆμα λέξεως)

Aristotle offers a cryptic defense of his taxonomy and its completeness:

> There is evidence of this [i.e., that these are the only six ways] both through induction and as a syllogism; if any other [syllogism] should be accepted there is also this one, that in just these many ways we might not signify the same thing by the same names and phrases.[6]

Presumably, the inductive evidence would consist in the inability to produce a false argument due to language that did not fit into one of the six classifications. The syllogistic evidence is less easy to reconstruct. We have here Aristotle's first general characterization of the common source of illusory arguments dependent on words: "not signifying the same thing by the same words and phrases." Perhaps, then, the syllogism that he has in mind would run like this.

> All failures to signify the same thing by the same names or phrases are due to these six phenomena.
> All illusory arguments due to language arise from failures to signify the same things by the same names and phrases.
>
> ---
>
> Therefore, all illusory arguments due to language arise from these six phenomena.

Kirwan[7] has claimed that the second premise is inconsistent with a later distinction that Aristotle makes *among* the types of fallacies due to language in *S.E.* 6. There he divides the six types into two subgroups. Homonymy, amphiboly, and Form of the Expression are due to double meaning (παρὰ τὸ διττόν) wherein the same name or phrase signifies more than one thing. But Composition, Division, and Accent are "due to there not being the same phrase, or the name being different."[8] In fact, there is no inconsistency here. The distinction between these two subsets of errors due to language will be detailed in this and the following two chapters. Briefly, errors of double meaning (i.e., homonymy, amphiboly, and Form of the Expression) occur when there is a failure to recognize that one and the same name (or phrase) signifies more than one thing. In the second group (i.e., Composition, Division, and Accent), this error is compounded by a prior failure to properly identify when there *is* one and the same name or phrase. As a result, in this latter group, *different* signifiers are mistakenly thought to be one and the same and, again, their different significations are overlooked. In both sets, the result is the same: a failure to signify the same things by the same names and phrases. In this and the next two chapters I examine Aristotle's treatment of each of these six sources of fallacies.

Homonymy

Homonymy in the *Categories*

Aristotle begins *Categories* 1 by writing:

> When things have only a name in common and the definition of being which corresponds to the name is different, they are called homonymous.[9]

Homonymy here is a relationship holding between things rather than between words. What makes two things homonymous is the fact that they are called by the same name, which has two definitions. Sometimes, instead of things being called homonymous, Aristotle says that names of things may be spoken homonymously.[10] This is not an inconsistency on Aristotle's part but a recognition that the homonymy that primarily pertains to things in the *Categories* also can be applied secondarily to the name possessed in common by those homonymous things. There is, however, one adjustment to be made when Aristotle uses "homonymy" to apply to names. It is no longer a relationship *among* names (as homonymy is properly a relationship among things). Rather, it applies simply to a name by virtue of that name's multiple signification. We might say that what it is to be the side of a river is *homonymous with* what it is to be a repository for money, but the word "bank" is simply homonymous. In more Aristotelian terms, homonymy according to the *Categories* is properly a πρός τι entity inhering in things. But when applied secondarily to the names of things, it loses that πρός τι status.[11]

Homonymy in *S.E.*

When we turn to *S.E.*, sophisms attributed to homonymy are sophisms due to names being homonymous. The *Categories* emphasis on things being homonymous is absent, yet it should be kept in mind that talk about homonymous names is only justified because the things so named are homonymous relative to each other. In *S.E.* 4, Aristotle offers three examples of fallacious arguments due to homonymy. The first argument is abbreviated but easily reconstructed from Plato's account of similar sophisms in his *Euthydemus*.[12] The homonymous word upon which the false refutation is constructed is the Greek μανθάνειν, to learn. The initial question is: Who is it that learns, those who know, or those who do not know? The answerer naturally would answer that those who do not yet know are the ones who μανθάνουσι. The apparent refutation occurs when the sophist secures concession to an apparent counterexample wherein there is both knowing and learning the same thing:

1. Those who know their letters are learning (μανθάνουσι) the things that are dictated to them.

   ———————

   Therefore, those who [already] know [something] are learning (μανθάνουσι) [that same thing].[13]

Two clarifications of the argument must be made in order to appreciate why the premise *is* a true counterexample to the initial claim, and why it is homonymy that creates the deception. First, the premise is *not* of the form: "those who know x, are learning y (by means of x)." Rather, the premise claims that "those who know x, are learning x," for the sophist is trying to refute the claim that "only those who do not yet know x are the ones who learn x." Second, in what sense are the letters already known the same as the things dictated that are being learned? Here a second fallacious argument is being assumed, namely, that "he who knows the letters knows the whole word, since the word is the same as the letters that compose it."[14] This particular error of identifying a linguistic whole with the sum of its linguistic components is an example of another linguistic error—Composition or Division—that will be discussed in chapter 4. For the present, though, it is enough to recognize that some reasoning of this sort allows the sophist to identify that which is already known with that which is being learned.

Aristotle explains the fallacy by claiming that the verb μανθάνειν is homonymous: it means both "to understand by making use of the knowledge" and "to acquire knowledge."[15] If we were to restate this in terms of the *Categories* concern about homonymous things, we would say that there are two types of activities, acquiring knowledge and using that knowledge, both of which can be called μανθάνειν in Greek. The Greek verb, then, is properly applicable both to the activity of the raw beginner being introduced to the principles of a discipline and to the advancing student deepening his or her understanding of those principles. Socrates explains this same distinction to Cleinias in the *Euthydemus*:

> [O]n the one hand, men apply "to learn" (μανθάνειν) to the sort of case when someone who at the beginning has no knowledge of some matter later obtains knowledge of it, and they call it the same thing when someone who already has the knowledge investigates by means of this same knowledge the same matter, whether in action or in speech. They usually call this latter activity "understanding" (συνιέναι) rather than "learning," but sometimes it is also called "learning."[16]

The second example of false reasoning due to homonymy is a complete argument.

2. Things that must be (τὰ δέοντα) are good.
   Evils must be (δέοντα).
   ———
   Therefore, evils are good.

Aristotle's explanation is that τὸ δέον signifies two things. It signifies something that is necessary (i.e., is inevitable), which is true of evils, and it signifies something that *ought* to be (i.e., is desirable), which is true of good things.

Aristotle next presents a pair of sophistical arguments attempting to show that the same individuals satisfy contrary descriptions.

3A.	The person who stood up is standing.	ὅσπερ ἀνίστατο, ἕστηκεν.
	The sitting person stood up.	ἀνίστατο ὁ καθήμενος.
	———	———
	The same person is both sitting and standing.	τὸν αὐτὸν καθῆσθαι καὶ ἑστάναι.[17]
3B.	The person who became healthy is healthy.	ὅσπερ ὑγιάζετο, ὑγιαίναι.
	The sick person became healthy.	ὑγιάζετο ὁ κάμνων.
	———	———
	The same person is both sick and healthy.	τὸν αὐτὸν κάμνειν καὶ ὑγιαίνειν.[18]

Aristotle explains that "the sick person" (or "the sitting person") sometimes signifies the one who is *now* sick and sometimes signifies the one who *was* sick before. Although Aristotle calls this a case of homonymy, we should note that "the sick person" differs in its multivocity from that of τὰ δέοντα and μανθάνειν. The latter two are examples of that use of homonymy from the *Categories*: things having a name in common but differing in definition. In contrast, "the sick person" does not signify two things of different definition, but signifies two different referents both having the same definition. What this fallacy illustrates is the "power of names" introduced in *S.E.* 1, whereby the same name signifies multiple particulars under a universal. In this case, the multiple particulars are only differentiated temporally. This mixing together of homonymous words based on double *sense* (homonymy as described in *Categories* 1) and homonymous words based on double *reference* (the power of names as described in *S.E.* 1) will reappear in Aristotle's examples of amphiboly.

Aristotle chooses his examples to show the two different ways in which homonymy may be abused in syllogistic reasoning. Double meaning is either in the premises or in the conclusion.

> Of refutations due to homonymy and amphiboly, some have more than one particular signification of the questions, while in others the conclusion is that which is said in many ways.[19]

In example (2) the two significations of τὸ δέον are exhibited, one in the first premise and the other in the second. The conclusion, however, is entirely univocal. In examples (1) and (3), the word with two significations occurs in the conclusion.

By these two different ways, all three types of sophistical appearances can be produced: namely, arguing from merely apparent endoxic premises, merely apparent arguing from real endoxa, and valid arguing to an only apparently relevant conclusion. When the homonymy is in the premises (i.e., in the middle term of the syllogism), validity is gained only if the middle term is read univocally. This, however, renders one or the other premise obviously false, and *a fortiori* nonendoxic.[20] If both premises are truly endoxic, there is only the appearance of a syllogism. When the homonymy is in the conclusion, the argument can be read either as a valid syllogism leading to an irrelevant conclusion or as a merely apparent syllogism. For example, in the argument about "the sick person," if the conclusion is understood to mean "the sick person before," the result is a perfectly valid syllogism leading to a true conclusion:

> The person who became healthy is [a] healthy [person now].
> The sick person [before] became healthy.
> ___________
> The sick person [before] is [a] healthy [person now].

Although a valid syllogism, the fact that sick persons later become healthy is irrelevant to the issue of whether a person can at one time be both sick and healthy. But if the conclusion is read as

The sick person [now] is [a] healthy [person now]

then the argument is only an apparent syllogism. Thus in all cases of homonymy, the sophistical appearance of valid reasoning is produced by a possible reading of the argument as valid. The price that one must pay for that reading is either an irrelevant conclusion or a false (nonendoxic) premise, depending upon whether the homonymy is in the conclusion or the premisses.

Amphiboly

It is time to consider Aristotle's examples of the second type of double meaning: amphiboly. There has been some dispute over the precise distinction that Aristotle wants to draw between homonymy and amphiboly. One attractive proposal is that homonymy is a semantic ambiguity, while amphiboly is a syntactic ambiguity. I believe that our modern distinction between semantics and syntax does indeed capture a large part of Aristotle's distinction. However, the former does remain our, and not his, distinction. Attempts to read the semantics-syntax distinction back into Aristotle are almost irresistible, so fundamental is that linguistic dichotomy in modern thought. However, Aristotle's thought here defies easy assimilation to the modern categories. Perhaps the closest Aristotle ever comes to distinguishing semantics from syntax is his distinction between homonymy and amphiboly, yet syntax alone cannot always account for his examples of amphiboly.

Owen and Hintikka have prudently avoided describing Aristotle's distinction in terms of semantics and syntax. Instead, they claim that Aristotle's distinction is between the ambiguity of a word (homonymy) and the ambiguity of a phrase or an expression (amphiboly).[21] I show below that this accurately accounts for Aristotle's examples. Irwin, however, has argued against this way of describing the distinction.[22] He claims that Aristotle's use of "amphiboly" covers both words and phrases. This disagreement with Owen and Hintikka stems from Irwin's emphasis on two passages outside of the *Organon* (*Rhet.* 1407a32, 37, and *Poet.* 1461a26) where, he claims, single words are labeled amphibolous.[23]

Hintikka and Owen note these latter passages but dismiss them as being exceptional.[24] In defense of the Hintikka–Owen position, however, I argue below that Aristotle's own explanation of the "amphibolous word" turns out to be dependent upon the ambiguous role the word can play within the context of the larger phrase. Owen does not mention the *Poetics* passage but interprets the *Rhetoric* passages also as referring to single words. He labels these uses "deviant." It does not seem clear to me that the *Rhetoric* texts require this concession to Irwin. However, before looking at the passages in dispute between Irwin, Owen, and Hintikka, we should examine the examples of amphiboly in *S.E.*

Amphiboly in *S.E.*

Aristotle offers five examples of double meaning due to amphiboly in *S.E.* 4. The first two examples are not cited within sophistical arguments. Instead, they illustrate the phenomenon that could be used to produce the false appearance of argument or refutation.

1. to wish me the enemy to capture
   τὸ βούλεσθαι λαβεῖν με τοὺς πολεμίους

This ambiguity might be understood as narrowly focused on the personal pronoun (με). Yet the pronoun is not homonymous, for it refers univocally to one and the same person. Rather, the ambiguity is located in the phrase λαβεῖν με τοὺς πολεμίους. It could mean that the enemy captures me, or that I capture the enemy. Within the phrase, the pronoun can operate either as the subject or object of the indirect phrase, and the fluidity of Greek word order allows either option.[25] The ambiguity is then syntactical, for the same well-formed Greek sentence admits two readings depending on the role one assigns to nouns and pronouns in indirect speech. Syntax alone, however, is not sufficient to produce the ambiguity. Were it not also for the rule that requires subjects in indirect speech to take the accusative case, the amphiboly could not exist as a well-formed Greek sentence.

2. Is it the case that whatever someone knows, that (he?) knows?
   ἆρ' ὅ τις γινώσκει, τοῦτο γινώσκει;

Again, τοῦτο can operate as either the subject or object in the second phrase. "Is it the case that whatever someone knows, that he knows?" Obviously yes. But reading the pronoun as the subject, we have the false claim that whatever someone knows, that thing itself knows something. In Aristotle's words, "For it is possible to signify either the knower or the thing known as if it were the knower in this expression."[26] Aristotle does not claim that the single word τοῦτο is ambiguous. Its antecedent is unequivocal: ὅ τις γινώσκει. He says instead that the problem is in the larger expression (τούτῳ τῷ λόγῳ). The phrase is ambiguous because of (1) the fact of Greek morphology, wherein nominative and accusative neuter pronouns have identical forms, and (2) because Greek word order allows either the subject or the object to precede the verb. Here, again, morphology and syntax team up to make a sophistical argument possible.

Aristotle's next example is quite similar to the preceding, though he gives the entire sophistical argument.

3. Whatever someone sees,	
(a) he sees that; (b) that thing sees.	ὃ ὁρᾷ τις, τοῦτο ὁρᾷ.
He sees a pillar.	ὁρᾷ δὲ τὸν κίονα.
———	———
[Therefore,] the pillar sees.	ὁρᾷ ὁ κίων.

Reading the first premise as (a) yields a true premise but an invalid argument. Reading it as (b) yields a valid argument but one having a premise that only appears endoxic but is actually an obvious falsehood. Greek syntax and morphology permit both readings for the first premise.

4. Whatever you claim to be,	
(a) that thing you claim is;	
(b) that you claim you are.	ὃ σὺ φῂς εἶναι, τοῦτο σὺ φῂς εἶναι.
You claim a stone to be.	φῂς δὲ λίθον εἶναι.
———	———
Therefore, you claim to be a stone.	σὺ ἄρα φῂς λίθος εἶναι.

Again, reading (a) yields a true premise but an invalid argument; (b) yields a false, nonendoxic premise but a valid argument. The amphiboly lies in the phrase τοῦτο σὺ φῂς εἶναι, where Greek syntax allows the τοῦτο to act either as the subject or the object of the infinitive.[27]

5. (a) Is it possible to speak of silent things?	
(b) Is it possible when being silent to speak?	ἆρ' ἔστι σιγῶντα λέγειν;

Aristotle states that the phrase σιγῶντα λέγειν can be understood in two ways, either as the one speaking being silent or as the things spoken about being silent.

What can we conclude from these five examples of amphiboly from *S.E.* 4? Examples (1) and (5) involve oblique contexts that set up the syntactic possibility of accusative subjects. Examples (2) and (3) also are syntactical confusions, abetted by the morphological identity of nominative and accusative neuter pronouns. It is that morphological ambiguity that permits the pronoun to play two different syntactical roles in the same sentence. Example (4) is the most complex, involving an oblique context as well as relying upon the syntactical rule according to which the subject of the main verb can be the unexpressed subject within the oblique context. In short, Greek syntax is not always alone sufficient for the generation of amphibolous phrases. Accidents of morphology sometimes are necessary for the appearance of the double meaning.

What is common to all of these examples is that they involve a word that signifies one thing but can play two syntactical roles within an expression. Often that syntactical ambiguity is made possible because of an accident of morphology. Because the word can play a double role in the larger context, Aristotle refers to the entire expression rather than to the single word as amphibolous. Do the examples of amphiboly outside of the *Topics* and *S.E.* agree with this summary? It is time to look at the disputed texts cited by Irwin. These are Aristotle's four references to amphiboly outside the *Organon*: one in the *Poetics* and three in the *Rhetoric*.

Amphiboly Outside the *Organon*

Poetics 25, 1461a25–26, is the text cited by Irwin and conceded by Hintikka as claiming that a single word is an amphiboly. I argue here that this is not the case. Part of the poetic art so prized among the Greeks included the ability to resolve the numerous "Homeric Questions." This was the ability to interpret Homer in such a way as to preserve his text from apparent contradiction or seeming nonsense. In *Poetics* 25, Aristotle offers six possible linguistic sources of these interpretive problems.[28] One of those sources is amphiboly, and Aristotle cites a line from the *Iliad* with the briefest of explanations:

> "more night had passed," for "more" (πλείω) is amphibolous.[29]

It certainly sounds as though Aristotle is calling the word πλείω amphibolous. However, contrary to the assumption of Hintikka and Irwin, this explanation turns out to be an elliptical reference to a larger Homeric phrase beginning with πλέω.

Aristotle has not quoted the entire Homeric sentence of dispute. In fact, he has not even included the part of the sentence that contains the interpretive problem. However, these Homeric passages had become such common grist for interpreters by the fourth century that the citation of the opening words was sufficient to communicate the entire problem.[30]

We happen, in fact, to possess Porphyry's account of Aristotle's detailed resolution of this problem.[31] In that account, the meanings of πλέω are never discussed. Instead, Aristotle is concerned with distinguishing the meaning of the subsequent genitive clause governed by πλέω. The two lines of the *Iliad* in question are:

> And now the stars had advanced, and more than two parts
> of the night had passed, and a third part still remained.[32]

The problem was how more than two parts of the night could pass and still leave a third yet to come. If *more* than two-thirds had passed, then there must be *less* than one-third left. The claim of those who read the single word πλέων as ambiguous is that Aristotle solves the difficulty by understanding it to mean "the greater part of" rather than "more than."[33] According to this solution, Homer means to say that the greater part of two parts of the night had passed, while one-third of the night remained. If such were the problem, Aristotle need only have distinguished between the two homonymous uses of πλέων. But Aristotle never discusses the senses of πλέων in his detailed resolution. Instead, he spends considerable time explaining that τῶν δύο μοιράων refers to the two equal halves of the night rather than to two equal thirds. This, of course, renders any reading of πλέων as "more than" arithmetically absurd. Now, instead of more than two-thirds of the night passing, we have more than two halves of the night passing! Aristotle's point is that Homer could only have meant that more than *one part* of the two equal halves had passed. That is, the genitive clause governed by πλέων must be read distributively, not collectively. Translating πλέων as "the greater part of" two halves is how English marks the genitive clause as distributive rather than collective. "The greater part of two halves" is equivalent to "more than one of the two halves." This is not an example of a single homonymous word. Rather, πλέων, plus a genitive phrase involving a numeral, is amphibolous.

Let me summarize here. In the *Poetics* reference to the Homeric Problem of *Iliad* 10, 252–53, Aristotle seems to say that πλέων is amphibolous. But in his detailed resolution of that problem preserved by Porphyry, he never discusses πλέων. Instead, he argues that the "two parts" mean two equal halves, and that the genitive phrase governed by πλέων can be read either collectively (producing the inconsistency with τρίτατη δ' ἔτι μοῖρα λέλειπται) or distributively (in which case, Homer's arithmetic works out). The amphiboly, then, is the phrase πλέων νὺξ τῶν δύο μοιράων, which Aristotle elliptically marks by citing the opening word. Here, as in the examples of amphiboly in *S.E.*, the amphiboly arises because some word can play two different roles within a particular phrase. In the Homeric verse, πλέων can play two different roles as it relates to the genitive clause τῶν δύο μοιράων, depending upon whether one understands the clause collectively or distributively.

Are the *Rhetoric* texts that mention amphiboly as "deviant" as Owen suggests? I discuss each of them in turn.

In *Rhetoric* I, 15, 1375b11, Aristotle is offering advice on the means of persuasion in forensic oratory. He says that if a law is amphibolous, then one must turn it about and see which way of taking it (ποτέραν τὴν ἀγωγήν) fits either justice or expediency and make use of that interpretation. Clearly the amphiboly is an entire sentence (or sentences) admitting of two different readings. Laws do not consist of single words.

In *Rhetoric* III, 18, 1419a20, the discussion is how to obtain the rhetorical advantage when interrogating one's opponents in public assemblies or law courts, or when answering questions posed by one's opponents.

> One must answer amphibolous questions by making distinctions in the expression but not concisely.[34]

Aristotle expands upon this point in *Topics* VIII:

> When [the question asked is one that] is said in many ways, it is not necessary to agree or to deny. . . . If the thing said is understood in many ways . . . and one knows in some cases how it is false and in some cases how it is true, one must indicate that it is said in many ways and that in one way it is false and in another way it is true. For if one should make the distinction later, it is unclear whether he also understood the amphibolous question in the beginning.[35]

There is no reason to understand amphibolous questions as deviating from the sense of "amphiboly" consistently maintained. Dialectical questions are not single words.[36]

In *Rhetoric* III, 5, 1407a32, 38, Aristotle is illustrating five principles of good Greek diction (ἀρχὴ τῆς λέξεως τὸ ἑλληνίζειν—1407a19–20). The third principle is not to use amphibolies:

> This is so unless one intends the contrary [i.e., intends to be ambiguous], as those do when they have nothing to say but pretend to say something. These sort of people say these things in verse, like Empedocles. For [verse] which goes on for a long time in circles tricks the listeners and they are affected just as the many are who listen to the seers. For when they [the seers] say amphibolous things, they [the listeners] express their assent—"Croesus, by crossing the Halys, will destroy a great realm."[37]

This instance of prophetic amphiboly remains a stock example in elementary logic texts. It is harder to situate into Aristotle's category of amphiboly, because neither morphology nor syntax seem to create a context of ambiguity. My suggestion is that the Croesus prophecy illustrates the "power of names" introduced in *S.E.* 1 as it applies to an entire proposition.

Consider again Aristotle's examples of misreasoning due to homonymy. These included cases not only of words signifying different kinds of things (μανθάνειν, τὰ δέοντα) but also words signifying the same kind of thing differently qualified (e.g., Socrates sick now, and Socrates sick before). The latter examples are more illustrative of the *S.E.* 1 power of names than of the narrow sense of homonymy defined in *Categories* 1. So too in the Croesus prophecy, this amphiboly exemplifies the power of common terms, when joined in a general predication, to signify multiple instances of that general predication. According to Herodotus (I, 91) the ambiguity in the oracle was which great realm (that of Cyrus or that of Croesus himself) would be destroyed. There is no ambiguous syntax here. This is a case where the prophecy makes a general claim that may signify more than one particular event. Aristotle says of cases like the Croesus prophecy that seers, in order to lessen their chances of being mistaken, "speak through the genera of the fact."[38] Aristotle next illustrates this relationship between the fact (τὸ πρᾶγμα) and its genus (τὸ γένος) by the relationship (1) between a particular number (πόσα) and the odd or the even numbers, and (2) between a particular future time (τὸ πότε) and the future in general.[39] What we have here, then, is a recurrence of that broad *S.E.* 1 power of names to signify many things, this time appearing in the context of amphiboly. It is not a single name that has multiple possible significations but an entire predication ("You will destroy a great realm") that can signify many particular events. Our inclination is to distinguish the multiple references of general names and the multiple instances of general predications from cases of real linguistic ambiguity. Aristotle, though, regards the multiple references of general predicates to issue in more examples of linguistic double meaning. That general power, however, cuts across the dichotomy between homonymy and amphiboly. If the fallacy is due to one general predicate signifying multiple particulars *regardless of context,* Aristotle classifies it as being due to homonymy. If the fallacy is due to general predicates in a propositional context signifying a general event that can be instantiated by multiple particular events, he classifies it as being due to amphiboly.

Problems with Aristotle's Distinction: The Argument of *S.E.* 17

The distinction just drawn between homonymous and amphibolous abuses of the power of common predicates is more easily made in theory than in practice. The problem is that all reasoning is propositional, and a homonymous term that applies to multiple particulars will always occur as part of a general predication. What is the difference between "You will destroy a great realm," where there may be many possible instances of "a great realm," and "The sick person is healthy," where there are many possible instances of "the sick person"? Can Aristotle justify the double meaning of "the sick person" in its fallacious setting as homonymous, while the double meaning of "a great realm" in its setting is amphibolous? I think not, and there is evidence that Aristotle himself had difficulty classifying fallacies based on multiple applications of common predicates as being either homonymous or amphibolous. That evidence comes from his discussion of an argument in *S.E.* 17, in which he identifies the same problem first as one of homonymy and later as one of amphiboly.

In *S.E.* 17, Aristotle offers some general strategy to avoid the appearance of being refuted by false arguments. He stresses that false arguments do not really refute but only appear to do so. Therefore, it is sometimes sufficient only to appear to resolve the false argument. This tactic has, I believe, been mistakenly interpreted by commentators such as Poste and Hamblin as a recommendation that one should fight sophistry with sophistry.[40] Such interpretations have contributed to the modern dismissal of Aristotelian dialectic as a serious philosophical tool. But nowhere here (or elsewhere) does Aristotle encourage sophistry. His point is that since the sophist has produced only the appearance of a refutation, then the philosopher's goal is to dispel that appearance. One can best accomplish this by resolving the fallacy outright. But sometimes one knows that the argument is fallacious, however, is unsure of the proper resolution. In such cases, it may be enough to cast suspicion upon the appearance of refutation by challenging the univocity of the premises. The assumption is, of course, that the premises truly are equivocal. Aristotle is careful to warn against leveling charges of ambiguity (whether amphiboly or homonymy) if there is a real refutation.[41] He does not countenance sophistical tricks to extricate oneself from a real refutation. The general thrust of his advice, then, is to embolden answerers to insist upon clear distinctions of meaning immediately upon being asked questions with ambiguities. Apparently, one sophistical tactic was to so overload the questions with ambiguities that if the answerer were to insist upon stopping to clear up all of them, he would strain the patience of the audience.[42] In order, then, not to alienate the listeners, answerers were tempted to overlook certain ambiguities only to discover themselves caught in an apparent refutation later on. Aristotle is conscious of this dilemma for the honest answerer who is trapped between the equivocating wiles of his sophistical challenger and the possible impatience of his listeners. Still, he recommends risking the irritation of others to put an end to the disingenuousness of the sophist. Aristotle offers an interesting analysis of one sophistical argument based upon multiple reference that shows his lack of sureness about how to classify it. The passage of interest reads as follows, with my emphases added:

> [A]
>
> If someone shall assume that there is a refutation *by homonymy*, in a certain way it will not be possible for the answerer to escape being refuted. For in the cases of visible things, it is necessary to deny a name which he said, and to affirm what he denied. For there is no benefit as certain people make the correction. For they do not say that Coriscus is musical and unmusical, but that *this* Coriscus is musical and *this* Coriscus is unmusical. For "this Coriscus" will be the same phrase as "this Coriscus" [who] is unmusical (or musical), which he both affirms and denies at the same time. But perhaps they do not signify the same thing (for neither did the names in the former case). As a result, what is the difference? And if in the one case he will allow one to say simply "Coriscus," and in the other case he will add "the certain" or "this," he commits an absurdity, for it belongs no more to one than to the other, for in whichever case it would make no difference.
>
> [B]
>
> Nevertheless, since it is not clear whether the one who has not distinguished the *amphiboly* has been refuted or not, and one is allowed to make distinctions in the expressions, it is clear that to grant the question while not making distinctions, but [granting it] simply, is an error.[43]

Aristotle is here describing a sophistical argument that he calls homonymous at 175b15 but amphibolous at 175b28–29. The sophistical conclusion is that Coriscus is both musical and unmusical. There might be two ways of appearing to reach that conclusion. If the premises involve Coriscus being at one time unmusical and at another time musical, we would have a fallacy similar to the *S.E.* 4 sophism that concluded "the sick person is healthy." That piece of false reasoning was attributed to the homonymous use of "the sick person" to signify either a person now sick or a person sick earlier. But such is not the reasoning here. In this argument, the premises involve two different persons named "Coriscus."[44]

When faced with an argument that involves "Coriscus" signifying two different individuals, Aristotle contrasts the wrong approach in [A] with the right approach in [B]. The wrong approach is to let the ambiguous use of "Coriscus" pass and to expect later to be able to answer an apparent refutation by drawing subsequent distinctions. According to Aristotle, if one does not immediately challenge the ambiguity of "Coriscus" (called homonymy in [A]), his later attempts to distinguish the two Coriscuses by different linguistic qualifications will be answered easily, and the audience will end up more sure than before that the answerer has been truly refuted. The right approach to the double Coriscus ambiguity (called amphiboly in [B]) is to challenge the univocity of the premises immediately, even before one knows what the intended refutation will look like, before one knows even whether the final refutation will turn out to be real or apparent. One should assume that any ambiguity in the premises might be leading to a false refutation and point out the ambiguity before the dialectician can even set up his argument. Aristotle acknowledges that merely questioning the premises does not constitute a resolution of the false argument, but if one treats such premises as though they were leading up to a false refutation, then by demanding clarification immediately, one is able to cast doubt in the minds of others on whether the intended refutation is real or not. By casting suspicion upon the premises at the outset, one sows seeds of doubt in the audience's mind as to the legitimacy of the eventual reasoning. Such a procedure manages to cast doubt upon the validity of the refutation and amounts to an *apparent* resolution. It is not a real resolution of the fallacy. For that, there must be more than an identification of the linguistic cause of the confusion.[45]

For our purposes, we need only note that Aristotle attributes the same fallacious argument leading to the conclusion that Coriscus is both musical and unmusical to homonymy and to amphiboly in the same text. This lapse, at odds with his insistence elsewhere of distinguishing between homonymy and amphiboly, might be explained by indecision over whether the signifier with multiple particular applications is the single word "Coriscus" or the expression "Coriscus is musical." Compounding the deceptiveness of this example is the use of an individual name having multiple referents. The understandable "error" assumed by the victim of this fallacy is that "Coriscus" signifies only one individual.[46] Aristotle's wavering over the classification of the "double Coriscus" argument does not shake his confidence in the distinction between homonymy and amphiboly. Throughout *S.E.* 17, he pairs the two as distinct types of the same kind of error: linguistic double meaning.[47] What it does show is that the clear distinction between homonymous and amphibolous double meanings when the referents are different *kinds* of things or different *kinds* of events is blurred when the referents are different particulars of the same kind.

CONCLUSION

In his account of fallacies due to double meaning, Aristotle incorporates two very different phenomena. One is the fact that some signifiers signify different kinds of particulars. The other is that all universal signifiers can apply to many different particulars under the same kind. Aristotle considers both phenomena to be cases of linguistic ambiguity.[48] The first phenomenon admits the clear distinction between (1) names possessing purely semantic ambiguity (i.e., independent of syntactical context), and (2) phrases possessing ambiguous meanings, because context fails to restrict the syntactical roles of particular words. In both cases, the words or phrases are universals covering multiple kinds of individuals.[49] This distinction covers most of Aristotle's examples of homonymy and amphiboly. However, Aristotle also has a concern about the second phenomenon: that general power of universals to apply to multiple individuals of the same kind. He tries to map that phenomenon onto the homonymy/amphiboly dichotomy, but it fails for the following two reasons. First, all of the examples (bottom row of following chart) are context-dependent and therefore amphibolous. But second, all involve universals that have, by definition, multiple signification, regardless of context, and they are therefore homonymous. Ambiguous references of universals to multiple particulars of the same kind simply fail to exhibit the sort of distinction that Aristotle wants to draw between homonymy and amphiboly.

Another difference between the two phenomena indicated by the two rows of the chart is their relative inevitability in natural language. Aristotle has already shown in *S.E.* 1 that the bottom row phenomenon—the power of names to signify universals and the multiple particulars under them—is inevitable, given the limited number of names and the unlimited number of things. Any language, designed as it is for the understanding of reality, will possess such a character. But there seems to be no comparable inevitability of those forms of ambiguity indicated in the top row. All of Aristotle's examples

	HOMONYMY Dual signification of words, context-independent	AMPHIBOLY Dual significations of phrases, word signification is context-dependent
Universals with multiple references to particulars of *different* kinds	μανθάνειν τὰ δέοντα	λαβεῖν με τοὺς πολεμίους τοῦτο γινώσκει τοῦτο ὁρᾷ τοῦτο συ φῂς εἶναι σιγῶντα λέγειν
Universals with multiple references to particulars of the *same* kind (= *Cat.* 1: synonymy) (= *S.E.* 1: power of names)	"the sitting person is standing" "the sick person is healthy" "you will destroy a great realm" "the double Coriscus"	

of amphiboly could be precluded by tightening certain syntactical rules and making some changes in morphology. And because names are tied to universal kinds, one should be able to eliminate homonymy by the creation of new names. Aristotle, however, shows no interest in such language reforms, and a more careful consideration of his beliefs about language goes some way toward explaining why. He believes in the conventional nature of language, whereby the names used to signify universals are determined by social agreement. He also believes in the inevitable power of those limited number of names to signify multiple particulars. The same conventional name, then, picks out several particulars, say *x* and *y*, because they are f, where f is the defining universal. But *x* and *y* will have more characteristics in common than just f, and different conversational contexts will make those different characteristics of *x* and *y* of primary interest. It is easy to see how the name properly signifying f and applying to *x* and *y* could gradually and by convention come to signify other kinds of universals with applications beyond just *x* and *y*. In short, standard names with the power to apply to multiple particulars of the *same* kind would naturally take on a range of significations of *different* kinds and thereby become homonymous in that strict sense of *Categories* 1 and *S.E.*[50] We might say, then, that although Aristotle's examples of the double meaning fallacies of homonymy and amphiboly regrettably mix together two quite distinct types of multivocity, there is a (perhaps inevitable) progression from one to the other in the natural development of ordinary language use.

Chapter 3

Form of the Expression

Introduction

The third form of false reasoning due to double meaning is that designated as resulting from the "Form of the Expression" (τὸ σχῆμα τῆς λέξεως). The phrase may have originated in a description of the new rhetorical style of Gorgias,[1] but Aristotle transforms the expression from a rhetorical term to a description of one of the sources of philosophical error. The examples of this fallacy are of somewhat more philosophical interest than the previous examples of homonymy and amphiboly. They include, for example, the important criticism of Platonic Forms known as the Third Man Argument. These errors are directly concerned with how words relate to things, and Aristotle believes that their resolution requires philosophical clarity about the distinctions among the kinds of things that exist. This belief that a thorough grasp of the different Categories[2] protects one against these sorts of errors suggests that most of these errors can best be described as "Category mistakes." I shall show, however, that some of Aristotle's examples do not involve inter-Categorial confusions and, contrary to Aristotle's suggestion,[3] knowing how to differentiate between Categories is not enough to render all such errors innocuous. The linguistic source of the persuasiveness of these errors stems from the false belief that similarities of word endings (whether noun declensions or verb conjugations) reflect analogous similarities in the things signified, or that similarities in word position in a sentence (i.e., its syntax) reflect analogous similarities in the things signified by other words in that same position. Aristotle rejects these two beliefs—that word position (syntax) and word termination (morphology) reflect specific ontological traits in the things signified—by appealing to his doctrine of the Categories. Ontological clarification is the key to resolving these linguistically based errors.

Aristotle provides important discussions of the error due to Form of the Expression at *S.E.* 4, 7, and 22. My analysis of the error begins with the *S.E.* 22 treatment of the resolution of the fallacy, because it provides the best general description of the error along with a wealth of examples, culminating in the Third Man Argument. A review of these examples reveals three variations on the general notion of a Category mistake. I then turn to the briefer *S.E.* 4 introduction to the fallacy. There Aristotle alludes to certain other examples, again generated by morphological similarities in word terminations, which seem to result in gender confusions. I compare these confusions with Aristotle's remarks on solecism elsewhere in *S.E.* I then show how such gender confusions fit within the threefold scheme of Category mistakes. Finally, I turn to *S.E.* 7, where Aristotle defends his claim that Form of the Expression is truly a fallacy due to language. Although this claim is uncontroversial, there is some disagreement over Aristotle's classifying this linguistic fallacy as being an instance of double meaning, such as homonymy and amphiboly. Kirwan (1979), for example, has argued that fallacies due to Form of the Expression are not proper examples of double meaning. I offer a partial defense of Aristotle against this charge but agree with Kirwan that Aristotle's taxonomy exhibits an unfortunate looseness in this instance.

FORM OF THE EXPRESSION AS A CATEGORY MISTAKE

Aristotle's general characterization of the fallacy due to Form of the Expression in *S.E.* 22 is that it is a confusion of a primary substance (τί ἐστι) with what is not a primary substance.

> It is clear also how we must meet those fallacies due to things not the same being said similarly, since we have the kinds of predications. For one person, when asked, granted that one of the things which signifies a substance does not belong [as an attribute], and another showed that some relative or quantity, which seems to signify a substance because of the expression, does belong.[4]

As illustrated by the examples that follow, the confusion is engendered by similar syntactical positions of names signifying things in different Categories. In particular, the error arises because

> we assume that each thing predicated of something is a certain this[5] (τόδε τι), and we give answer as if it were one thing.[6]

In the following sections I analyze Aristotle's examples of fallacies due to Form of the Expression. Most of his descriptions of these examples are highly abbreviated. In my English renderings I have expanded many of them into a dialectical format for greater clarity and appreciation of the false refutation. Although I discuss each example from *S.E.* 22, I do not follow Aristotle's order.

Confusion of Substance with Quantity

Example 1 "If a person first has something, then later does not have it, did he lose it?"
"Yes."
"But a person first has ten dice, then later, by losing one die, he does not have ten dice. But it is not the case that he lost what he first had and later did not have, namely, ten dice."

Aristotle answers this argument by saying that it confuses *what* someone lost with *how much* or *how many* someone lost, since things in both Categories can be signified by words in the same predicate position.

> Although asking the question about what (ὅ) he has, his conclusion is about how many (ὅσα), for ten is how many. If then he asked from the start "has someone lost as many things as he had earlier but now does not have?" no one would grant it, but instead [he would grant] "either that many or some part of them."[7]

Example 2 "Is what someone knows, known either by learning it from someone or by discovering it for oneself?"
"Yes."
"But take the case of a pair of things, one of which was learned and the other was discovered. It is false that *that pair* is known either by learning or by discovering."

This apparent refutation assumes that the disjunction between learning and discovering is understood exclusively. Whatever is signified by "what someone knows" is either learned or discovered, but not both. Aristotle answers this sophism by saying that although it is true to say of each thing one knows that it was either learned or discovered, it is not true of all the things one knows that they were either *all* learned or *all* discovered.[8] By including these errors among the fallacies of double meaning, Aristotle is noting that "what someone knows" (ὅ τις οἶδεν) can apply to single items of knowledge or to a number of items of knowledge. The fallacy does not arise because of what the phrase signifies, but because of how many things the phrase signifies. This differentiates the example from cases of amphiboly, such as the "great realm" prophecy. In both cases, context underdetermines which particular item or event the universal expression might signify. The mistake of Croesus was to confuse one particular (realm) with another particular (realm). But the victim of the "what someone knows" fallacy confuses one particular (item of knowledge) with a quantity of particulars. This mistake of treating many individuals as though they were a single individual may not seem to us an example of ambiguity, but it illustrates the wide application of "ambiguity" in Aristotle to cover universal expressions that admit multiple applications. Because the fallacy hinges on confusing a quantity with a single particular, Aristotle regards it as a kind of Category mistake resulting from the Form of the Expression.

Confusion of Substance with Relative

Example 3 "Is it possible to give away what one does not have?"
"No."
"But someone having ten dice can give away only one, even though he does not have only one."

Aristotle explains this fallacy:

> For "only" does not signify a particular thing, or some quality, or some quantity, but how someone has it relative to something, i.e., that he does not have it with another.[9]

In other words, I cannot give away X if I do not have X, but I can give away only X even if I do not have only X. The term "only X" plays the same syntactical role as any term signifying an individual. But it does not, thereby, signify an individual. It signifies, says Aristotle, a relationship among individuals.

Other variations on the signification of "only" as πρός τι include:

Example 4 "Could someone strike with a hand that he does not have?"
"No."
"But he could strike with only one hand, though he does not have only one hand."[10]

Example 5 "Could someone see with an eye that he does not have?"
"No."
"But he could see with only one eye, though he does not have only one eye."[11]

Example 6 "Is it not the case that what one person gives to another, that second person has?"
"Yes."
"But one person gave another only one vote, yet the second person has ten votes, not only one."[12]

It is not just "only" that leads to false reasoning when it is thought to signify a substance because of its syntactical position. This sort of confusion arises with any adverb in a predicative position, as shown by other sophistic counterexamples to the truth that one cannot give away what one does not have:

Example 7 It is possible to give away [something?][13] quickly what one did not have quickly. This is fallacious, says Aristotle, "for to give quickly is not some thing (τόδε) but is to give in some manner (ὧδε)."[14]

Example 8 It is possible to give away painfully what one had with pleasure.[15]

Aristotle's final example of an adverb misunderstood as signifying a substance is:

Example 9 "That which a learner learns, is that not what he learns?"
"Of course."
"But it is possible for him to learn something slow quickly, so what he learns ("something slow") is not what he learns ("something quickly")."[16]

As Aristotle says: "It is not, then, *what* he learns but *how* he learns that was said."[17]

All of these examples confuse *what* something is (ὅ or τόδε) with *how* or *in what manner* something is (ὡς or ὧδε). Aristotle seems to assimilate all of these to confusions of substance with relative, but is every qualification of manner to be considered an example of πρός τι? Elsewhere Aristotle reserves confusions of unqualified predication (ἁπλῶς) with qualified predication (ἢ πῄ ἢ ποὺ ἢ πὼς ἢ πρός τι)[18] to the extralinguistic fallacy of *Secundum Quid.* In chapter 8 I discuss the relationship between that form of fallacy and these examples classified under Form of the Expression. For now, I merely note that what is mistaken because of the Form of the Expression for a substance is sometimes difficult to place in any of the Categories of things that are.[19]

Confusion of Substance with Quality

Example 10 "Did x write what is written (ὃ γέγραπται) by x?"
"Yes."
"But what is written is the false statement 'You are sitting,' and what x wrote was the true statement 'You are sitting.' (For it was true when x wrote it.) Therefore, if what is written by x is the same as what x wrote, then what is written is both true and false."[20]

Aristotle's analysis of this argument involves an interesting claim about truth-values. He claims that "what is written" (ὃ γέγραπται) is the statement "You are sitting," independent of the truth-value of the statement.

> [This refutation is fallacious] for that a statement or opinion is false or true signifies not a particular substance but a quality [of that statement or opinion]. For the same account applies also in the case of an opinion.[21]

Aristotle believes that truth-values belong to statements as qualities belong to substances.[22] Claims such as "a true statement was written" or "a false statement was written" predicate passivities (i.e., attributes of being written) of ontological substances, but the grammatical subjects ("true/false statements") signify ontological composites: statements with the qualities of truth or falsity. Like the double meaning between substance and quantity in the phrase "what one knows" (ὅ τις οἶδεν), here there is a composite reference to substance and quality in "what is written."

Confusion of Substance with Time

Example 11 "Does someone tread on that through which he walks?"
"Yes."
"But he walks through the whole day, although he does not tread on the whole day."

The source of the confusion, again, is that the Greek signification for walking through the whole day (βαδίζει τὴν ἡμέραν ὅλην) is expressed by putting the time walked in the same syntactical position and case as something walked on. "But it was not what he walked through, but when he walked that was said."[23]

All of the above examples of this fallacy type involve the confusion of substance with nonsubstance because of syntactical similarities between their respective signifiers. In the next examples, morphological similarities between signifiers result in confusions among the nonsubstance Categories.

Confusion of Activity with "Being-Affected"

Example 12 "Is it possible both to be doing and to have done the same thing at the same time?"
"No."
"But surely it is possible to be seeing and to have seen the same thing at the same time in the same respect."[24]

According to Aristotle,

> if someone in that situation, having granted that it is not possible to be doing and to have done the same thing at the same time, should say that it is possible to be seeing and to have seen [the same thing at the same time], he is not yet refuted if he should say that seeing is not a doing but a being-affected.[25]

The reason for the false belief that seeing is an activity is that it has the same morphological terminations as do other verbs signifying activities.[26] This is clear from Aristotle's next example.

Example 13 "Is there some example of being-affected that is also some activity?"
"No."
"Are not 'he is cut,' 'he is burnt,' and 'he is perceiving' said in the same way, and all signify something being-affected? And again, 'to say,' 'to run,' and 'to see' are said similarly to each other. However to see certainly is to be perceiving, so that it is at the same time both an activity and being-affected."[27]

Here the English fails to capture the linguistic similarities in the Greek verbs. Whereas seeing is a species of perceiving (and therefore is in the same Category), the Greek verb "to see" is grammatically active, but the Greek verb "to perceive" is grammatically deponent (and therefore has only passive forms). This is a case of similarity in Greek word morphology inducing a belief in ontological (i.e., Categorial) similarity among the things signified. A similar confusion is exemplified in the *S.E.* 4 introduction to the fallacy to which I now turn.

Confusion of Activity with Quality

> False refutations due to the Form of the Expression occur when what is not the same is expressed in a similar way, for example . . . when some quality is expressed like a quantity or some quantity like a quality, or some activity like a being-affected or some condition like an activity, and the others as distinguished earlier. For it is possible to signify what is not an action as some action by the expression. For example, "being healthy" [an active verb] is said similarly in form of expression to "cutting" or "housebuilding" [also active verbs]. Nevertheless, the former is a certain quality and somehow signifies a condition, while the latter is a certain action. The same way, too, in the other cases.[28]

Here the similarity in morphological terminations of the active verb "to be healthy" and other verbs signifying activities induces the false belief that being healthy is an activity instead of a quality.

FORM OF THE EXPRESSION FALLACIES THAT ARE NOT CATEGORY MISTAKES

All of the examples examined so far involve inter-Categorial confusions. Either nonsubstances are thought to be substances because of the similar syntactical position of their signifiers, or something in one nonsubstance Category is thought to be something in a different nonsubstance Category because of the similar morphological endings of their signifiers. If these exhausted Aristotle's illustrations of the fallacy type, we could be satisfied with the general description of the error as a Category mistake, but Aristotle also includes several examples that involve no inter-Categorial confusion.

Confusion of a Particular with a Universal

The single most important philosophical example of a fallacy due to Form of the Expression is the notorious Academic argument of the Third Man against the Platonic doctrine of separated Forms. The error here, and its cause, says Aristotle, is

> that there is some third man besides [man] itself and the individual men. For "man" and each common name signifies not a certain this (τόδε τι), but a certain kind (τοιόνδε τι), whether of quantity, or relative, or one of those sorts [of Categories].[29]

In *Cat.* 5, Aristotle warned against treating secondary substances as primary substances. Even there he places the blame for such a confusion on "the form of the name."

> But in the case of the secondary substances, though it appears from the form of the name—when one speaks of man or animal—that a secondary substance likewise signifies a certain this, this is not really true; rather, it signifies a certain qualification, for the subject is not one, as the primary substance is, but man and animal are said of many things.[30]

The "certain qualification" (ποιόν τι) is an unfortunate description of these universals, for it makes them sound like members of the Category of Quality. Aristotle goes on immediately to try to deflect such a possible confusion, but he is hindered by his choice of the same word (ποιόν) for a *kind* of substance as he uses for a *quality* of a substance.[31] In the *S.E.* passage, Aristotle is more fortunate in his description: the universal word, he says, signifies not a quality but a kind (τοιόνδε τι), and such kinds can exist in any Category.[32] Aristotle concludes:

> It is evident, then, that one must not grant that what is predicated in common over all [the members of a class] is a certain this, but signifies instead either a quality or relative or quantity or one of these kinds.[33]

Confusion of One Particular Substance with Another

Aristotle, at one point, illustrates the error of Form of the Expression by citing what seems to be no more than metonymy:

> nor does "to drink the cup" [signify] what he drinks but from what he drinks.[34]

Certainly that "from which" liquid is drunk is as much a substance as the liquid that is drunk. It is simply a different substance. This confusion, then, is not inter-Categorial. It is a confusion between different items in the same Category.

Confusions Based on Gender Terminations

Aristotle's remaining examples of errors due to Form of the Expression in *S.E.* 4 are best understood on the model of the example of "to drink the cup":

confusions of different items within the same Category. They raise, however, enough difficulties of interpretation on their own to warrant a fairly extended analysis.

> [The error arises when], for example, what is male is expressed like a female, or what is female like a male, or a neuter (τὸ μεταξύ) like one or the other of them.[35]

One difficulty in understanding this passage arises from the lack of any Greek grammatical apparatus to distinguish word use from mention.[36] Exacerbating the use-mention confusion is the additional fact that words such as τὸ ἄρρεν and τὸ θῆλυ can refer either to male and female natures or to masculine and feminine grammatical genders. As a result, one might understand the fallacy under consideration to arise from mere grammatical irregularities. The issue in this passage, however, is not that of the relationship of one word to another word with respect to their grammatical forms. Rather, the confusion results from extralinguistic objects with masculine or feminine natures being signified by words having noncorresponding grammatical genders. In my English translations, ἄρρεν and θῆλυ are translated as "masculine (word)" and "feminine (word)" when I interpret them to be signifying grammatical gender, and as "male" and "female" when signifying nonlinguistic natures.

The ancients were interested in both issues: the consistency of grammatical forms among words of the same gender, and the consistency of the grammatical gender with the extralinguistic nature. These two issues, however, need to be distinguished if we are to appreciate Aristotle's comments on this particular source of false reasoning. The former issue arises from the empirical observation of grammatical irregularities in Greek conjugations and declensions. Ancient grammarians and Hellenistic editors of earlier Greek texts believed that this phenomenon required an explanation, and a famous controversy was joined between the "analogists" and the "anomalists" to account for grammatical irregularities.[37] This is *not* the problem that Aristotle has in mind when he discusses errors due to the Form of the Expression. Aristotle's fallacy is concerned with the underlying "word-thing" relationships rather than the "word-word" relationships that dominated grammatical discussions in the Hellenistic period.

There is evidence both from Aristophanes and in Aristotle that within the Socratic and sophistic movements there was a belief that Greek word gender was somehow connected to the sexual nature of the thing signified. In the *Clouds*, Aristophanes shows Socrates correcting Strepsiades' use of nouns of standard masculine declensions to refer to objects with feminine natures. Socrates coins a feminine form of ἀλεκτρυών (cock) to refer to a hen, and he changes the masculine form κάρδοπος (kneading-trough) into a feminine to better reflect the nature of the object (660 ff.).[38] Aristotle cites Protagoras as claiming that because μῆνις (wrath) and πήληξ (helmet) are both masculine *things*, it is incorrect Greek to decline them as feminine words, which is what they in fact are. Protagoras inveighs against Homer who opens the *Iliad* by describing wrath as οὐλομένη (accursed, feminine) rather than οὐλόμενος (accursed, masculine).[39]

Aristotle, in claiming that such correlations of noun termination with sexual nature produce false reasoning, is clearly distancing himself from Protagoras. Yet in the *Rhetoric,* Aristotle refers to a similar-sounding Protagorean project with approval. A close examination of the relevant texts (*Rhet.* III, 5, and *S.E.* 14) shows that Aristotle's judgments of Protagoras are consistent. He approves of the Protagorean program for the avoidance of solecism but withholds his approval of the further step of correlating word morphology to ontological natures.

Before looking at those two texts, a brief summary of the preceding is in order. Aristotle's examples of false reasoning due to Form of the Expression appear in three variations.

1. Most are true Category mistakes. These involve inter-Categorial confusions (e.g., substance with quality, substance with quantity, substance with relative, substance with time, activity with being-affected, activity with quality), arising either because of syntactical or morphological similarities between signifiers.
2. One involves the confusion of a universal with a particular in the same Category (the Third Man Argument).
3. One involves the confusion of a particular in one Category with a different particular in the same Category (the metonymous example, "to drink the cup").

Into which of these variants do confusions of the sexual natures of things fall? The likeliest candidate is the third—a substance of one sex is confused with another substance of a different sex. Aristotle, apart from several notorious passages in the *Politics,*[40] shows little interest in the sexual distinction among animals apart from its biological role in species generation. Sexual distinctions within species are considered quite unimportant, especially when measured against the importance of the distinctions between species. Aristotle is clear though that sexual genders belong to animals not as mere qualities or non-necessary attributes of substances. Confusion of a male with a female is not, then, an inter-Categorial error. Rather, it is the case that male or female belongs καθ' αὑτό to an animal in the same way that odd or even belongs to number, and equal or unequal belongs to a quantity.[41] In other words, sexual gender belongs to an animal, because what it is to be an animal is part of the definition of being male or being female.[42] These so-called "*per se* 2 predicates," regarded as mutually exclusive and jointly exhaustive disjunctive predicates,[43] are neither species differentiae nor non-necessary accidents. Confusing an animal possessing one of the disjunctive properties with an animal possessing another of the disjunctive properties amounts to confusing different particular substances.

FORM OF THE EXPRESSION AND SOLECISM: ARISTOTLE AND PROTAGORAS

An examination of the relationship between fallacies due to Form of the Expression and instances of solecism is important for two reasons. First, it makes clear the distinction between Aristotle's and Protagoras' approach to conventional language use. Second, by situating fallacies due to Form of the Expression as ontological confusions, not just grammatical confusions (solecisms), it raises anew the question of why these are called by Aristotle fallacies "due to language." One answer that this section forestalls is that linguistic fallacies are those errors that linguistic clarification is sufficient to correct. The answer that part 2 will develop is that they are errors whose correction necessitates some linguistic clarification. But more is required than just that.

In *Rhet.* III, 5, Aristotle lists five rules for speaking proper Greek. One of them, we saw earlier, was the avoidance of amphibolies, which was illustrated by the Croesus oracle. Aristotle continues:

> Fourthly, [to distinguish] as Protagoras distinguished the kinds of names, male and female and inanimate. For it is necessary to render these correctly, "She, having come and conversed, departed."[44]

Whatever it is that Protagoras is said to have done here, Aristotle approves of it. We know Protagoras believed that modifier words should agree in gender with the nature of what they modify, regardless of the grammatical gender of that modifying word.[45] The presumption here, then, is that this threefold division of male, female, and inanimate, is concerned as much with things as with words. Such a presumption is furthered by the choice of σκεύη for the third classification rather than either τὸ μεταξύ or οὐδέτερα, which became the usual terms to signify grammatically neuter words from the fourth century onward.[46] However, Protagoras cannot be concerned solely with natures, for Aristotle calls what are being distinguished "the kinds of names." At issue here seems to be the *names* given to males, females, and inanimates. The analysis of this passage by Cope provides the clearest explanation of what Protagoras advocated.[47] Particular names of persons, animals, gods, and so on (i.e., anything that admits sexual classification) must be distinguished from the names of inanimate things (regardless of their grammatical gender) and grammatical agreement maintained with the sexual *kind* of the antecedent. In the example that Aristotle cites ("She, having come and conversed, departed"), the insistence would be upon using the feminine forms of the participles if the antecedent was female. To do otherwise would be to fall into solecism, the subject of *S.E.* 14.

Solecism (σολοικισμός) is one of the five aims of public debate that Aristotle lists in *S.E.* 3. These five Aristotle arranges in order of desirability, with the best result being the refutation (ἔλεγχος) of the opposing position (i.e., reasoning to the contrary of the opponent's claim). If one cannot refute the opponent's claim, Aristotle lists four other inferior ways to cast suspicion upon the credibility of the opponent's position. He devotes three chapters (12 through 14) to these four inferior goals of public dispute. According to *S.E.* 3, the second best thing to refuting the conclusion of one's opponent is to show him holding some other false position in the course of the examination. Thirdly, if one cannot expose any falsity in the opponent's claims, one should try to expose some paradox or highly doubtful claim. Fourthly, if that fails, one should attempt to reduce his opponent to solecism, which Aristotle says is the use of a grammatically incorrect ("barbarous") expression.[48] The last resort is to reduce the answerer to babbling (ἀδολεσχῆσαι), which Aristotle glosses as "saying the same thing many times."[49]

All of these are legitimate goals in debate, although they, like refutation proper, can be sophistically abused. That is, there can be apparent but not real falsehoods, paradoxes, solecisms, and babblings. In the chapter on solecism, Aristotle's first task is to illustrate the possibility of merely apparent ungrammaticality. He does this by appealing to certain things Protagoras "used to say":

> It is possible [1] to produce [a solecism], [2] to appear to do so when not doing it, and [3] to not seem to do so when doing it, just as Protagoras used to say, if wrath and helmet are male things. For the one who calls [wrath] "accursed" [feminine form] commits a solecism according to him [i.e., Protagoras], but does not appear to do so to the rest, and the one who calls it "accursed" [masculine form] appears to but does not commit a solecism.[50]

It needs to be noted that Aristotle is careful *not* to commit himself to the Protagorean example, even while offering it as illustrative of the sort of thing he means by an apparent-but-not-real or a real-but-not-apparent solecism. The contrast drawn in the example is between how one was to speak proper Greek *according to Protagoras* (κατ' ἐκεῖνον) and how this sort of speech appeared *to everyone else* (τοῖς ἄλλοις). There is no reason to think that Aristotle would not consider himself in the latter class.[51]

Having exemplified what an apparent solecism might look like (and thus the possibility of such a thing), Aristotle generalizes that such an effect could be produced by art (τέχνη), thereby opening up the danger of systematic sophistical abuse. In particular, the Greek neuter pronoun has a grammatical feature that lends itself to sophistical exploitation. Neuter pronouns fail to discriminate between masculine or feminine referents. Sometimes it is grammatically acceptable for neuter pronouns to stand in for masculine or feminine names. This often occurs when one is unsure of the sex of the pronoun's referent. However, when the referent is known to be a male or a female, avoidance of solecism requires the appropriate masculine or feminine pronoun.

> Virtually all apparent solecisms are due to this, when the inflected ending fails to signify that the word is masculine or feminine, but that it is neuter. For while "he" (οὗτος) signifies a masculine word and "she" (αὕτη) signifies a feminine word, "this" or "it" (τοῦτο) is meant to signify the neuter but often it signifies each of the other genders, for example, "What is it?" (τί τοῦτο;) "Calliope" or "wood" or "Coriscus."[52]

Aristotle is saying that responses such as τοῦτο ἔστι Καλλιόπη and τοῦτο ἔστι Κορίσκος are apparent-but-not-real solecisms.[53]

One crucial difference between Aristotle and Protagoras becomes evident from their respective examples of apparent-but-not-real solecisms. For Protagoras, the appearance of solecism arises when common usage is violated for the sake of linguistic reform (e.g., treating "wrath" as masculine). It is common usage itself that is really solecistic. For Protagoras, even inanimate things may possess masculine or feminine natures, and the language needs to be reformed to reflect such realities. Such changes result in apparent-but-not-real solecisms—failures of grammatical agreement which, however, turn out to be real agreements of nature. For Aristotle, on the other hand, it is common usage that is only apparently solecistic. Aristotle turns out to be the defender of anomalies in the general rule of gender agreement when those anomalies are sanctioned by ordinary language. His advice in *Rhet.* III, 5 is the relatively tame recommendation to maintain grammatical gender agreement with any antecedent that has a sexual nature. But he has not, it seems, taken the more radical step of Protagoras in assigning sexual natures to inanimate objects and demanding the use of word terminations signalling those natures, even if the result sounds barbarous.[54]

Aristotle also acknowledges the possibility of real-but-not-apparent solecisms. He provides no examples of these, inasmuch as the production of these is of little dialectical value to the sophist, who seeks appearances, with or without any corresponding reality. Forcing one's opponent into an unrecognized solecism, even if real, is cold comfort in a dialectical contest.[55]

Aristotle's conclusion to the *S.E.* 14 treatment of solecism returns us to the original issue of fallacies due to the Form of the Expression.

> And in a certain way a solecism is similar to those refutations which are called "due to the same things not [expressed] similarly." For just as in those cases [one happens to be deceived] with respect to things, so in these cases one happens to speak incorrectly with respect to names. For man and white are both things and names.[56]

Aristotle here extends the literal sense of "solecism" (ἐπὶ τῶν ὀνομάτων) in order to produce the metaphor of committing a solecism with respect to things (ἐπὶ τῶν πραγμάτων). This is what the fallacy due to Form of the Expression is. It is one thing to be led by morphological or syntactical similarities into ungrammaticality. The error amounts to a violation of proper speech—a sin against linguistic convention. It is another thing to be led by morphological or syntactical similarities into a Category mistake or some intra-Categorial confusion. These errors amount to violations of the way things are—sins, so to speak, against the natural order. Both phenomena are linguistically generated. For Aristotle, however, the error of solecism never leaves the domain of language.[57] Errors due to Form of the Expression leap out of the domain of language, because clarification about the relationship of words to the things they signify is alone not sufficient to rectify the error. One also must know how to distinguish among the different kinds of things signified.

Form of the Expression As a Linguistic Fallacy of Double Meaning

Aristotle has taken great pains to emphasize that the error which double meaning engenders is one concerned with things, not just with words. Such was the point of the contrast drawn with solecism. He is equally emphatic in pointing out that although the errors engendered involve things, their source lies in the words used to signify those things.

> For this reason, too, this manner [of reasoning] is counted among the fallacies due to language, first of all since the deception arises more often in the investigations with others than by oneself (for the investigation with others is through words (διὰ λόγων), but the investigation by oneself is no less through the thing itself (δι' αὐτοῦ

> τοῦ πράγματος). Further, it is possible to be deceived even by oneself when one conducts the investigation on the basis of the word. Again, the deception arises from the likeness, and the likeness from the language.[58]

The contrast between λόγος and πρᾶγμα here cannot be that between word and external object.[59] Solitary inquiries can no more be conducted by means of external objects than can joint inquiries. Aristotle seems to have in mind here those intermediaries between spoken words and external objects, namely, the "affections of the soul" referred to in *de Int.* 1. Aristotle's characterization of such affections is never very detailed.[60] Two important characteristics, however, are that they are "likenesses of actual things (πράγματα)," and unlike human speech, they are the same for all persons.[61] For our purposes we may simply refer to them as concepts and accept Aristotle's dubious assumption that reasoning through concepts is somehow possible nonlinguistically.[62] According to Aristotle, to the extent that one reasons about things through concepts, he is removed from the influence of the spoken words that signify those things. It is the grammatical similarities among those spoken sounds that make the verbal expressions deceptive.

So far so good as an explanation for why this is a fallacy due to language. But in what sense are errors generated by Form of the Expression akin to errors generated by homonymy and amphiboly in being examples of double meaning? If "double meaning" is understood as the same word or phrase signifying more than one nonlinguistic entity, it is easy to see how fallacies due to homonymy and amphiboly fall into this subgroup. It is less easy to account for fallacies due to Form of the Expression. Kirwan summarizes the problem this way:

> The sophist who tries it [i.e., Form of the Expression] on is not using one expression with two meanings, and his inexpert respondent does not infer identity of meaning from identity of expression. Rather, he infers identity of meaning-type from identity of expression-type. Form of expression misleads when it does not match *form* of meaning.[63]

Kirwan here offers a neat characterization of this type of fallacy: identity of meaning-type inferred from identity of expression-type. Such a description does contrast with that of double meaning as it appeared in fallacies due to homonymy and amphiboly: identity of meaning inferred from identity of expression. To accept Kirwan's characterization of these fallacies we must allow a generous extension of application for "meaning-types" and "expression-types." Consider what Kirwan's "meaning-types" must encompass. They must signify: (1) the different Categories of things; (2) the difference between universals and particulars in each Category; and (3) the difference between particulars in the same Category. The same is true of Kirwan's rendering of the source of the error: identity of expression-types. To cover all of Aristotle's examples, an "expression-type" must include different words having identical terminations and different words in the same syntactical position. But in the latter case, confusing universal for particular predications in the same Category (e.g., Third Man problems) involves the same expression (not the same expression-type) having two different significations. How then does this sort of fallacy due to Form of the Expression differ from fallacies due to amphiboly?

The difficulty in strictly demarcating each Form of Expression fallacy from the other double meaning types is, once again, that Aristotle includes multiple applications of universals as a form of multivocity. Someone may be led to believe that "Socrates is different from a man" by confusing it with an expression of the same form such as "Socrates is different from Coriscus." But that just amounts to being ignorant that "man" signifies both a universal and many particulars. Is it an error due to Form of the Expression? Why not call it an error due to amphiboly, such as "You will destroy a great realm"? Or is it due to homonymy such as "The sick person is healthy"? The problem, I suggest, is that Aristotle fails to recognize that the "power of names" that he identified in *S.E.* 1 as a kind of multiple signification is distinct from three other forms of double meaning:

1. the same name having semantic double meaning independent of syntactical context (homonymy);
2. the same phrase having semantic double meaning because of constituent names having double syntactical roles (amphiboly); and
3. names with the same terminations or same syntactical positions signifying different kinds of things (Form of the Expression).

If we regard (3) as the purified class of errors due to Form of the Expression, we may accept Kirwan's description of these errors as "inferences from identity of expression-type to identity of meaning-type." What are we to say of Kirwan's charge that (3) does not precisely fit Aristotle's description of double meaning errors ("the same thing not being signified by the *same* names or phrases")? Kirwan is correct: "one die" and "only one die" are not the same phrase signifying different things. In Aristotle's defense, however, he does acknowledge a range of uses for the concept "sameness." Two verbs having identical terminations or two nouns occupying identical syntactical positions in a sentence may be "the same" in different ways from the sameness of, for example, the same word (τὰ δέοντα) signifying things that must occur and things that ought to occur. As Aristotle notes in *S.E.* 7, deceptions due to homonymy and amphiboly arise because of people's inability to distinguish the different ways things are said, and among the more difficult cases to distinguish he cites "one and being and the same."[64] In the last of these, Aristotle indicates in the *Topics* that things are said to be the same in roughly three ways: the same in number, the same in species, and the same in genus.[65] Elsewhere he presents alternative and additional ways in which things are the same.[66] In *Meta.* I, 3, he distinguishes being the same without qualification (ἁπλῶς) from being the same in some respect, which renders things "similar" (ὅμοια).[67] Aristotle's inclusion of fallacies that stem from "the similarity of the expression" (διὰ τὴν ὁμοιότητα τῆς λέξεως) under the general description of "the same thing not being signified by the same names and phrases" reveals perhaps a certain looseness in his terminology. Still it is a correctable looseness in light of his more careful attempts to disambiguate notions of sameness. Any likeness is a real, albeit not unqualified, instance of sameness.

This employment of the multiple uses of "the same" as a defense of the inclusion of fallacies due to Form of the Expression among the errors of double meaning is, however, costly, for it appeals to a distended sense of "the same" that appears so overly inclusive as to admit a qualified sameness to virtually any two items. In the end, even after purifying the class of errors due to Form of the Expression to true inter-Categorial confusions, Kirwan's question remains: Why should these be understood as errors of double meaning on a par with errors due to homonymy and amphiboly?

The full answer must wait until part 2 on Resolutions. There it will be shown that the one characteristic that unifies these diverse types of error and distinguishes them from the other types of double meaning errors listed earlier is that they are resolved in the same way. It turns out that our trying to understand Aristotle's typology by isolating a different effect common to each set of examples is the wrong approach. In fact, it reverses Aristotle's method of codification. Examples of false reasoning are unified into groups and differentiated into subgroups not by the nature of the confusions that they engender but by the different ways that confusion is cleared up. In part 2, this characteristically Aristotelian method will be set forth in detail.

Chapter 4

Composition, Division, and Accent

Difficulties and Procedure

The analysis of Aristotle's fallacies due to Composition (παρὰ τὴν σύνθεσιν) and Division (παρὰ τὴν διαίρεσιν) is complicated by three related problems. First, there is a modern tradition of labeling errors based upon parts-whole confusions as errors of Composition and Division. According to this interpretation, an error due to Composition may arise by predicating what is true of each part of x to the whole of x, and conversely an error due to Division may arise by predicating what is true of the whole of x to some part of x. The warrant for such an analysis is often traced back to Aristotle's description and examples of the fallacies in his treatment of fallacious enthymemes in *Rhet.* II, 24. Unfortunately, such an analysis fits almost none of the examples Aristotle provides in *S.E.* This raises the second problem: how are we to relate the parallel accounts of Composition and Division in *S.E.* and *Rhetoric*? In particular, and this is the third problem, does Aristotle intend to signify by his terms "Composition" and "Division" two distinct fallacy types (*S.E.*), or are these two alternative names for one and the same type of false reasoning (*Rhetoric*)?

I will speak briefly about the latter two problems. There is no scholarly consensus on the relationship between the detailed account of false reasoning in *S.E.* and the briefer treatment of false enthymemes in *Rhet.* II, 24.[1] Much of the terminology is the same, but the two treatments elude any complete harmonization.[2] Attempts to uncover some deep consistency between the two accounts should not interfere with a close analysis of the texts on their own terms. Aristotle's treatment of these errors in *S.E.* is sufficiently detailed and illustrated not to require commentary on what may be only superficially similar examples in the *Rhetoric*. Accordingly, I intend to reserve most of my analysis of the *Rhetoric* material to "Confusing Linguistic Parts and Wholes" and "C/D Fallacies in the *Rhetoric*" below. I do, however, make one important use of the *Rhetoric* material to shed light on *S.E.* In the *Rhetoric,* Aristotle treats Composition and Division as a single fallacy. I claim that this procedure is preferable to his listing them as two different types in *S.E.* Nowhere in *S.E.* does he offer any justification for the distinction, nor does he provide any hint of how someone distinguishes an example of one type from an example of the other type. In part 2 on Resolutions, I show that elsewhere even in *S.E.* Aristotle suggests that the two are actually one fallacy type.

Finally, what can we say about the recent tendency to regard Composition and Division as parts-whole confusions? In the first place, what is cited as the strongest Aristotelian affirmation of such an interpretation is based on a mistranslation.[3] In the Revised Oxford translation, *Rhet.* II, 24, 1401a25–26, is rendered: "Another [fallacious] line is to assert of the whole what is true of the parts, or of the parts what is true of the whole." This cannot be defended as a reading of the Greek text. Aristotle's description of this fallacy is "to say [something] by combining what has been divided or [to say something] by dividing what has been combined."[4] Only by a premature interpretive decision can "what has been divided" (τὸ διῃρημένον) and "what has been combined" (τὸ συγκείμενον) be related as "parts" and "whole." Both Greek participles refer to a composite unit, but how Aristotle intends to relate the composites signified by the two descriptions must await clarification from his examples of the fallacy. I show below that the *S.E.* examples preclude understanding that relationship as one of extralinguistic parts to their wholes. Secondly, what tends to be ignored in the parts-whole interpretation of the fallacy is Aristotle's positioning of the fallacy among those "due to language." Fallacies due to Composition and Division, then, insofar as they are παρὰ τὴν λέξιν, have something to do with the way words are arranged.[5] In contrast, the problem with most parts-whole confusions is that the parts and the whole at issue are extralinguistic.[6] Aristotle is well aware of such extralinguistic confusions. He insists that certain philosophical errors about change, for example, be resolved by acknowledging that physical beings are composites, and that there are truths predicable of the composite that cannot be truly predicated of both form and matter (e.g., composites are generated and undergo corruption).[7] Such parts-whole confusions are never described as errors due to Composition or Division.

There is, however, a special sort of parts-whole confusion that Aristotle does include among the fallacies of Composition and Division. When the parts and whole at issue are themselves linguistic units and someone mistakenly believes that what is true of the components of a proposition (e.g., its phonemes, letters, or words) is true of the entire proposition, Aristotle does classify such a fallacy as one of Composition or Division. There are, I believe, two possible examples of such an error in *S.E.* Because these are exceptional cases to Aristotle's general account of Composition and Division in *S.E*, I reserve discussion of them to "Confusing Linguistic Parts and Wholes" below. There I also justify Aristotle's inclusion of them as fallacies of Composition or Division, despite their untypical nature.

The first procedural principle to be followed in analyzing the *S.E.* treatment of Composition and Division, then, is to expect the generating error to involve a confusion between different arrangements of the words making up the component propositions of the refutation. The second principle, generally overlooked by commentators, is that, among the linguistically based fallacies, Aristotle associates Composition and Division more closely with the fallacy of Accent (προσῳδία) than with fallacies of double meaning.

> Of fallacies in language, some are due to double meaning, e.g., homonymy, and the sentence [amphiboly], and the likeness of form (for it is customary to signify all things as a certain this), but Composition and Division and Accent are due to the sentence not being the same or to the different name.[8]

Aristotle's distinction here is easy to misread and crucial to get right. One might be tempted to suppose that linguistic fallacies *not* due to double meaning must involve signifiers *without* multiple signification. After all, the characteristic cause of double meaning errors, we have seen, is the same name or sentence signifying different things. But Aristotle does not say this. He says that linguistic fallacies not due to double meaning arise when the name or sentence is not the same. The distinction is, perhaps, poorly labeled, for it has nothing to do with the multiplicity of things signified. It has to do with the multiplicity of the signifier itself. Whereas linguistic fallacies of double meaning involve the same name or sentence signifying different things, linguistic fallacies not due to double meaning involve different names or sentences signifying different things. But why would anyone suppose that different names or sentences did *not* signify different things? The error arises because the speaker or reader mistakes the different names or sentences for the same. There is, then, a compound error at work in fallacies of Composition, Division, and Accent. Not only is there a mistake about the things being signified (mistaking different things for the same), but also there is the prior mistake about the signifying words or phrases themselves (mistaking different signifiers for the same). Part of the resolution of these fallacies involves knowing what constitutes a distinct linguistic signifier.

Finally, these two principles, that the error originates in the language and that it involves a confusion of signifiers, contribute to a third curiosity. By being based on confusions of word arrangement, they would seem to be akin to errors of amphiboly. But by being distinguished from fallacies of double meaning, they must be importantly different from amphibolies. As a result, Aristotle concedes the appearance of amphiboly in errors of Composition and Division while denying that amphibolous double meanings are actually at work.[9]

We have, then, three important clues to analyzing the fallacies of Composition and Division: (1) They are generated by the arrangement of words in a sentence; (2) they are closely related to fallacies due to Accent; and (3) they must not be confused with amphibolies. Our starting point, then, will be Aristotle's analysis of the fallacy due to Accent. Therein lies an important hermeneutical clue for an understanding of the various examples of Composition and Division in *S.E.*: the distinction between written and spoken argumentation.

FALLACIES DUE TO ACCENT

The one point that Aristotle stresses about fallacies due to Accent is that they are generally restricted to written accounts and poetry rather than to oral dialectic.[10] Given the relative novelty of written accounts of dialectical argumentation even in the fourth century, it is not surprising that Aristotle's two examples in *S.E.* 4 are both taken from Homer. Neither seems to have anything to do with argumentation within the text but rather are concerned with argumentation about the text.

> For example, some people make corrections in Homer in opposition to those who refute [Homer's text] as absurd when he said "part of which [i.e., οὗ] decays in the rain." For they resolve it by means of the accent, saying the ου more sharply [i.e., as οὐ].[11]

Any ambiguity in Homer's text here is exclusively a function of the written transcription. In oral recitation, Homer's line would be unambiguous. But in the Greek alphabetic transcription, which lacked the subsequent conventions of accent and breathing marks, the line would be ambiguous, and one reading would be vulnerable to the charge of absurdity.

In the second example, the charge against Homer seems to have been theological. He might be accused of attributing to Zeus deceitful behavior unbecoming the god. The context is the sending of the Dream to Agamemnon in *Iliad* II in order to rouse him to attack the Trojans. The written text (though not the spoken recitation) of Homer could have been understood in two ways. Either Zeus is actively deceiving Agamemnon by saying to the Dream "we grant [δίδομεν] him the fulfillment of his prayer," or he is giving to the Dream the command "to grant [διδόμεν = διδόναι] him the fulfillment of his prayer." Such a fine grammatical distinction was thought presumably to shift the moral culpability away from "the father of gods and men" to the Dream.[12]

Although illustrating the problem of reading an unaccented text, the Homeric examples do not themselves embody fallacious arguments. They are texts about which disputes trading on Accent had arisen. In *S.E.* 21, Aristotle concedes that not only are there no spoken fallacies due to Accent, but there are no actual written ones either. He does proceed to invent one that hinges on the same written ambiguity of the Greek ου that was seen in the first of the Homeric examples, but Aristotle makes it clear that this is an example devised for the occasion and not an actual sophism that has arisen within the eristic tradition. That tradition is an oral tradition, but this sophism could only arise in writing. The fact that Aristotle flags his example in this manner shows that his study of false reasoning should account not just for actual cases of false oral argument but possible cases of false argument in transcription. The artificial argument runs like this[13]:

> "Is it not the case that a house is where you lodge?"
>
> (ἆρά γ' ἐστὶ τὸ οὗ καταλύεις οἰκία;)
>
> "Yes."
>
> "And is it not the case that 'you do not lodge' is the denial of 'you do lodge'?"
>
> (οὐκοῦν τὸ "οὐ καταλύεις" τοῦ "καταλύεις" ἀπόφασις;)
>
> "Yes."
>
> "But you said that a house is where you lodge. Therefore a house is a denial."
>
> (ἔφησας δ' εἶναι τὸ οὗ καταλύεις οἰκίαν· ἡ οἰκία ἄρα ἀπόφασις.)

There is in this contrivance a flagrant use-mention confusion, but Aristotle's point is plain: in the absence of accent and breathing notations, the written form of ου καταλυεις is ambiguous. It is not a case, however, of amphibolous double meaning, because it is not the same phrase signifying two different things but two different phrases. Aristotle is here a witness that, even up to the middle of the fourth century, oral expression had priority over writing.[14]

We may summarize our findings in this manner. If, in a fallacious argument, a word or phrase, as pronounced, signifies more than one thing, it is a fallacy due to double meaning. But when *differently* pronounced words or phrases are transcribable by the identical set of alphabetic symbols, and that transcription plays a role in a fallacious argument, then there is a fallacy due to language but not due to double meaning. Aristotle confirms this in *S.E.* 20:

> Although in writing the name is the same whenever it is written from the same letters and in the same way (even though people already are producing additional signs [παρασήμα]), the pronunciations are not the same.[15]

It may be that one of the reasons for the absence of any written sophisms due to Accent was this fourth-century innovation of providing παρασήμα: distinguishing marks to show pronunciation.[16] What does seem evident is that with oral dialectic beginning to be transcribed for a reading public, a complete accounting of linguistic fallacies would need to encompass two phenomena: fallacies of double meaning in which the word or phrase *as spoken* is ambiguous, and fallacies in which the spoken word or phrase is unambiguous, but the transcription is ambiguous. Aristotle's division among linguistic fallacies between those due to double meaning and those not due to double meaning is, I suggest, a reflection of the gap between the refinements possible in speech for significant utterance and the absence of such refinements in fourth-century Greek transcription. Certainly this is the case with fallacies due to Accent. It remains to be seen how this hypothesis applies to the examples of Composition and Division.[17]

FALLACIES DUE TO COMPOSITION AND DIVISION (C/D)

C/D Fallacies Are Not Examples of Double Meaning

Among the scattered examples of false reasoning due to Composition and Division (C/D) in *S.E.* and *Rhetoric*, some are duplicates and some involve only minor variations on one another.[18] I begin with a set of examples that I describe as errors of participial Composition, because they involve participle constructions whose transcriptions might be read in different ways. In *S.E.* 4, Aristotle writes:

> These sorts of arguments are fallacies due to Composition. For example, its being possible to walk when sitting and to write when not writing (for it does not signify the same thing if someone should speak when dividing [the phrase] and when combining [the phrase] as if it were possible to walk-when-sitting. And likewise if someone should combine the phrase "to-write-when-not-writing." For it signifies that he has the power to write-when-not-writing; but if he

> does not combine, it signifies that when he is not writing he has the power to write.[19]

What is it that appears ambiguous about τὸ δύνασθαι καθήμενον βαδίζειν? As transcribed, one may read the participle as though the expression signified "although sitting, he is able to walk," or as "he is able to walk-while-sitting." The latter reading is, of course, absurd. Aristotle, however, does not consider this an amphiboly. To understand why, we must recall his claim that fallacies due to C/D are related to fallacies due to Accent. Aristotle's point seems to be that speakers of particular Greek phrases can signify different things by the way they vocally combine or separate the verbal components of those phrases. Presumably, this would be done by variations in vocal pauses and emphases. Such vocal pauses can be reflected in written English translations only by artificial marks (e.g., the use of hyphens in my translation above). Whereas it is possible to-write when not-writing, it is not possible to write-when-not-writing. So, too, in Greek transcription there are no natural markings to distinguish the two different significations. As a result, this fallacy appears amphibolous in writing but could only arise in conversation through a change in the manner of oral delivery of the same word sequence. If the answerer claims that it is impossible to write and not write at the same time, the sophist gains the concession that it is possible for someone to write when he is not writing (understood in its normal uncombined sense, i.e., the possibility of writing can be simultaneous with actually not writing). He then refutes the original claim by repeating the concession *verbatim*, but with the same sequence of words differently pronounced to produce a different combination — "so it is possible to-write-when-not-writing" — where the possibility applies to the entire predicate of "writing-when-not-writing." In this way, the fallacy *can* arise orally, but it does so by using a different signifying sentence, although composed of the same words. Whether such a change in the signifying expression actually confounds a *listener* would depend upon whether that listener knew how to individuate signifying expressions. On the other hand, a fourth-century B.C. Greek reader of such an argument would be much more prone to deception. Since Greek writing omitted all breaks between words, not to mention any punctuation within a sentence, the reader had the double task of first deciphering word divisions and then making the proper connections between words to distinguish proper phrases with all of their proper modifiers. Just as παρασήματα such as accent or breathing marks could go a long way toward removing nonoral deceptions due to Accent, so, too, symbols such as commas could remove much of the nonoral ambiguity found in fallacies due to C/D. The mere technological absence of written devices for signaling vocal pauses and variations in pronunciation renders fallacies due to C/D a much more present danger to the reader of dialectic than to the oral participant. Still, the oral participant could be fooled if he did not realize that signifying phrases are differentiated by not only the phonemes spoken but by the way they are spoken. In sum, then, the Greek phrase τὸ δύνασθαι καθήμενον βαδίζειν is not an amphiboly, because as spoken it has only one signification. Only when written can it seem that the same string of words signifies two things. The two ways of vocalizing it constitute two different signifiers, not one signifier with a double signification.

We have here, then, two examples of C/D ("walking when sitting" and "writing when not writing") that seem to fit our hypothesis about the predominantly nonoral nature of these linguistic fallacies. There is, however, another approach to these examples that has found favor among some modern discussions of the fallacy. It is worth a look at this approach to see how it differs from Aristotle's analysis, and why Aristotle would reject it. According to William and Martha Kneale, the distinction behind the example is between two meanings of "possibility": absolute (ἁπλῶς) possibility and relative possibility. The idea is that what might be absolutely possible for someone (e.g., not writing) is not possible relative to certain other conditions (e.g., when writing). Relative possibility in this case, then, is possibility temporally qualified. The Kneales write that

> what is possible in itself need not be so in relation to all other statements. In both Greek and English the same words are used to express both absolute and relative necessity and possibility.[20]

Put that way, it sounds as though the Kneales are citing an equivocation — the homonymous use of δύνασθαι—as the cause of the fallacy.[21]

Aristotle's first answer to the Kneales probably would be to point out the distinction between fallacies due to linguistic double meaning and fallacies due to *Secundum Quid.* When one uses a word in a relative sense, elliptically and without mention of the pertinent qualifications, so that it might be mistakenly understood in an absolute sense, not only is that not homonymy but for Aristotle it is not even an error due to language. Confusions between absolute (ἁπλῶς) and relative (πρός τι) possibility are confusions outside of language; they are examples of the error of *Secundum Quid.* But merely to cite the reclassification is not to answer the Kneales' charge. Aristotle still must defend the extralinguistic nature of *Secundum Quid* fallacies. In chapter 8 I argue that this task is not without its difficulties.

The second and stronger answer to the Kneales occurs in *S.E.* 20, where Aristotle treats the resolutions of fallacies. I discuss the architectonic role of resolutions in the makeup of Aristotle's taxonomy of false reasoning in part 2. A look now at that reply to the Kneales' proposal will serve as a preview to the part 2 analysis.

In *S.E.* 20, another example of participial C/D is presented, along with a rejection of a Kneales' type of analysis. Aristotle presents the following sophism:

> Might you not do those things which you are able to do and in the way you are able to do them? Although you are not playing the harp, you have the power to play the harp. Therefore you might play the harp when not playing the harp.[22]

As in the earlier examples of participial Composition, Aristotle is calling attention to two ways of vocally combining the words of the conclusion. While one might be able to harp when not-harping, one cannot harp-when-not-harping.[23] But some people resolve the sophism differently, says Aristotle. Although conceding the first premise, "You can do those things which you are able to do and in the way you are able to do them," they deny the sophist's conclusion that it is possible to-harp-when-not-harping:

> For [they say that] they did not concede that he will do absolutely (πάντως) that which he is able to do. Being able to do it and being able to do it absolutely (πάντως) are not the same.[24]

This is very close to the Kneales' distinction between relative and absolute possibility, where the ἁπλῶς ability to do x is the ability to do x πάντως. Aristotle insists that this is not a proper resolution, because it cannot be used to resolve other sophisms of the same type that do not involve participles.[25] For instance, another example of the sophism of C/D is based upon the position of the temporal adverb "now" (νῦν) with a perfect verb. The illustration comes from *S.E.* 20, 177b20–22.

> Is it true to say right now that you have been born? Therefore [it is true to say that] you have been born right now.
>
> ἆρ' ἀληθὲς εἰπεῖν νῦν ὅτι σὺ γέγονας; γέγονας ἄρα νῦν.

The problem is that νῦν might be taken as temporally modifying the speaking (εἰπεῖν) or the being born (γέγονας). What makes this an example of Composition rather than amphiboly is that (a) ἀληθὲς εἰπεῖν ὅτι "γέγονας" νῦν is a different signifying phrase from (b) ἀληθὲς εἰπεῖν ὅτι "γέγονας νῦν". The use in our transcription of quotation marks to distinguish between the two phrases reflects the different ways of vocalizing the two expressions.

Another factor that might contribute to the seeming validity of the move from (a), which is simply a restatement of the uncontroversial premise, to (b) is that *present tensed* verbs do admit the inference. So, for instance, from (a') "It is true to say 'you are living' now," one can infer (b') "It is true to say 'you are living now'." But one cannot make the inference with future or past actions. It may be true to say "you will graduate" now, but that is not to imply that "you will graduate now."

The Primacy of Oral Speech

What unites the previous sophism with the examples of participial C/D is made explicit in another example that involves neither a case of participial C/D nor an awkwardly placed temporal modifier. Yet it confirms the central role in all of these examples of how one pronounces a particular string of words. The sophistical arguer begins by obtaining concession to the following:

> That with which (ᾧ) you saw someone being beaten, is it not that (τούτῳ) with which he was beaten? And that with which (τούτῳ) he was beaten, did you not see him [being beaten] with that (τούτῳ)?[26]

As long as the antecedents of ᾧ and τούτῳ offer specifications of the object of sight, the two questions express mere tautologies. But because the antecedents also might signify the means of sight, how one chooses to combine the words in speech makes a world of difference.

> For it is not the same to [say that I] see "with-my-eyes-someone-being-beaten" [= "someone who is being beaten with my eyes"] and to say that I "see-with-my-eyes" someone being beaten.[27]

Aristotle's point is that the written Greek, like the written English "to see with my eyes someone being beaten," records two different signifying expressions, depending on where one chooses to pause in the vocalization. In his explanation of this fallacy, Aristotle admits the appearance of amphiboly in the original question. But he then denies that there is any more double meaning in the word sequence of the sophist's question than there is in the letter sequence ορος. The latter is in fact two words, either "mountain" (ὄρος) or "boundary" (ὅρος), depending upon the pronunciation.[28] The principle that is crucial for an understanding of Aristotle's examples of Composition and Division is then stated:

> Although in writing the name is the same whenever it is written from the same letters (στοιχεῖα) and in the same way . . . the pronunciations[29] (τὰ φθεγγόμενα) are not the same. As a result, fallacies due to Division are not cases of double meaning.[30]

It is these pronunciations, these spoken sounds, that are the primary bearers of linguistic significance. The same written στοιχεῖα do not suffice to individuate the primary (i.e., spoken) linguistic signifiers. This follows both for the written στοιχεῖα of words (i.e., letters) and for the written στοιχεῖα of entire sentences (i.e., words or phrases). There can arise from the same transcription of the στοιχεῖα of spoken *words* fallacies of Accent. And from the same transcription of the στοιχεῖα of spoken *sentences* can come fallacies of Composition and Division. In neither case is there double meaning of one and the same linguistic signifier.

Further Examples

In this section I show that seven of the remaining nine *S.E.* illustrations of fallacies due to C/D exemplify the gap between the signifying role of enunciation and that of written transcription. I discuss the two exceptions in the following section.

In *S.E.* 4, Aristotle offers four examples of Division. They follow the same pattern as his examples of Composition. Neither of the first two is set within the context of an argument. Each is an independent quotation, susceptible to two different interpretations, depending upon vocalization.

Example 1 I have made you a slave being free.[31]

The false utterance is the claim that I have made you "a free slave" (δοῦλον ὄντ' ἐλεύθερον). By a vocal pause after δοῦλον, however, the same words should be understood to claim "I have made you, who up to now have been free, a slave."[32]

The written ambiguity of the second example is impossible to reproduce in English.

Example 2 The divine Achilles left fifty men one hundred.[33]

Two different things could be signified, depending upon where, or whether, one divided the phrase πεντήκοντ' ἀνδρῶν ἑκατὸν. If one joined "men" to the following "one hundred," the claim is that Achilles left one-half of his men ("fifty of the one hundred men"). If one divided "men" from the "one hundred," the claim becomes the absurdity that Achilles left twice as many men as he had ("one hundred of the fifty men").[34]

The other two examples of Division are of a mathematical nature:

Examples 3 and 4 "Since five is two and three, it is also both even and odd, and the greater is equal [to the lesser] for it is just that much and yet more besides."[35]

In the first case, the difference is between "five is two-and-three" and "five is two, and [it is] three." The second argument makes the same false inference: from the fact that (x + y) = (x) + (y), one infers that (x + y) = x and (x + y) = y (i.e., the greater is equal to the lesser). The difference, again, is between (x + y) being x-and-y, and (x + y) being x, and [also being] y. It bears repeating that Aristotle does not consider these to be cases of double meaning. The two word sequences, insofar as they are pronounced with different pauses, amount to two different signifiers. The failure to recognize that how the sentence is spoken is part of Aristotle's criterion for individuating significant sentences causes some modern analysts to account for these examples by appeals to double meaning. This particular argument, for instance, has been analyzed as due to an equivocation on "and" (καί) between its signification as a conjunction and its signification as a mathematical functor for addition.[36] But this is thoroughly non-Aristotelian. A word such as καί cannot possibly have a double signification, because it fails to have *any* signification.[37]

There are two remaining illustrations of Composition from *S.E.* 20: "the good cobbler" and "the knowledge of evil" arguments.

Example 5 Is it possible for a good man who is a cobbler to be bad? But someone might be good who is a bad cobbler, so that a good cobbler will be bad.[38]

The conclusion appears in writing to be amphibolous, reading the σκυτεύς either with ἀγαθός or with μοχθηρός. The false conclusion arises only in the first case, when "cobbler" is joined to "good." On the second reading, we have the unobjectionable restatement of the previous premise: "a good man will be a bad cobbler." These, however, would be for Aristotle two utterances, depending upon how the words are vocally combined. The drawing of the sophist's conclusion results only if the listener fails to distinguish the different spoken signifiers.[39]

The next example in *S.E.* is comparable to the two earlier examples of Division: "five is two and three" and "the greater is equal to the lesser."

Example 6 Things the knowledge of which is good, are they not good things to learn? But the knowledge of evil is good. Therefore, evil is a good thing to learn. But surely evil is both evil and something to learn, so that evil is an evil thing to learn. But the knowledge of evils is good."[40]

The first argument of this dialectical exchange is not fallacious. It leads to the agreed-upon conclusion that evil is a good subject to know about (σπουδαῖον μάθημα τὸ κακόν), because all μαθήματα are good and τὸ κακόν is a μάθημα. The second argument leads to the contrary claim (κακὸν μάθημα τὸ κακόν), because τὸ κακόν is both a μάθημα and κακόν. The fallacy of Composition occurs in this counterargument. Since evil is an evil, and it is something to learn (with the division between the two predicates), it is an-evil-to-learn (the predicates combined). The inference from divided to combined predicates is just the reverse of the inference from the combined predication of "five is two-and-three" to its divided form, "five is two, and it is three."

The final *S.E.* 4 illustration of Composition admits several interpretations. I offer one that most closely follows the pattern seen in the other examples of this fallacy. There is no single English translation that captures both readings of the Greek transcription. I present the two readings as separate translations.

Example 7 (A) Being able to carry one thing alone is to be able to carry many things.

(τὸ <u>ἓν μόνον</u> | δυνάμενον φέρειν πολλὰ δύνασθαι φέρειν.)

(B) Only being able to carry one thing is to be able to carry many things.

(τὸ ἕν | <u>μόνον δυνάμενον</u> φέρειν πολλὰ δύνασθαι φέρειν.)[41]

In the first reading, μόνον is combined as an adjective with ἕν and separated from δυνάμενον. It makes the reasonable claim that an ability to carry one thing alone is consistent with an ability to carry many things. In the second reading, μόνον is combined as an adverb with δυνάμενον and separated from ἓν. It makes the absurd claim that having only one ability is consistent with having some other ability. The transcription warrants both readings, though the vocalization will make one or the other claim. Two points are important to remember. First, this is not a homonymous use of μόνον between its signification as an adverb ("only") or as an adjective ("alone"). It is only its position in the larger context of τὸ ἓν μόνον δυνάμενον φέρειν that permits the grammatical double role. Second, this is not amphibolous, because each vocalization of the entire sentence yields a single unambiguous sense.[42]

With the advent of extensive systems of punctuation (and such systems were an inevitable development as languages adapted to cultures more dependent on written records than on oral speech), the probability of anyone committing Aristotle's fallacies of Composition and Division became exceedingly remote. Transcription mechanisms in modern European languages have brought the written phrase closer and closer to the spoken phrase. Today, confusions based upon C/D find their place generally in popular comedy. When a humorist offers the couplet

> Time flies like a bird;
> Fruit flies like a banana.

he obtains his comic effect by intentionally blurring the different vocalizations of the second line. By joining "fruit" with "flies" and pausing before "like," one signifies an unremarkable fact about certain insects. But by pausing after "fruit," one signifies a humorous absurdity. The English transcription admits both readings.

Aristotle's fallacies of C/D stand as monuments to the changes in Greek orthography beginning in the fourth century. By the end of the third century those changes had become established. According to the *Grammar* of Dionysius Thrax (second century B.C.), the first task of grammatical studies was learning how to properly read texts aloud (ἀνάγνωσις). This included: (1) attention to the style of delivery; (2) mastery of the features of prosody, which included attention to accents, length of syllables, breathings, and internal sentence punctuation; and (3) correct separation (διαστολή) of the utterance, from which one obtains the "overall sense" (περιεχόμενον νοῦν). Skill at this separation meant knowing when to elide words and when to pronounce a word sequence without a pause (ὑφέν) and with pauses (ὑποδιαστολή).[43] By Dionysius' day, Greek orthography had caught up to Greek pronunciation. There was little room left for fallacies of C/D to appear.

Confusing Linguistic Parts and Wholes

I have tried to show that Aristotle's fallacies due to Composition and Division are not generally confusions of parts with wholes. They are confusions of one significant utterance with a different significant utterance, where the distinction lies in how one vocally combines and separates the various component words. There is a variant of this fallacy, however, that does qualify as a parts-whole confusion. It arises when the parts and whole in question are linguistic units themselves. If one assumes that everything true of all of the parts of a significant utterance is true also of the whole utterance, then if it is true that one knows what all of the component parts of an utterance signify, one might conclude that one knows what the entire utterance signifies. Plato introduces such a sophism in the *Euthydemus*, where knowledge of all of the letters of the alphabet is falsely thought to entail knowledge of everything signified by any combination of those letters.[44]

In both the *Rhetoric* and *S.E.*, Aristotle alludes to this eristic tradition. Among the examples is the fallacy

> regarding the one who understands the letters (στοιχεῖα), that he knows what is spoken (τὸ ἔπος). For what is spoken is the same [as the letters].[45]

In *S.E.* 4, 166a30-31, Aristotle offers an example of C/D that has baffled commentators and led to several textual emendations. All we are given is a single conditional sentence. Reconstructions of that conditional sentence attempt to allow for two different ways the sentence can be read, depending upon how one combines and divides the words. I suggest, however, that it may not be the conditional sentence itself that is supposed to admit two readings. Aristotle may be abbreviating the same argument about linguistic parts and wholes from the *Euthydemus*. The best manuscripts read the line in question as μανθάνει νῦν γράμματα εἴπερ ἐμάνθανεν ἃ ἐπίσταται. Among the four principal modern English translators, only Forster (Loeb) retains this reading and tries to make sense of it. Ross, Pickard-Cambridge (Oxford), and Poste all offer their own emendations.

Before examining these four treatments, we should emphasize one point. The sophism cannot rest upon the homonymy of μανθάνειν between "learning" and "making use of what is learnt." However we are to read the fallacy, it must not equivocate on any single word.

Forster's reading

Forster's reading is initially attractive on two counts. It retains the received text, and it does suggest two ways of vocalizing the same string of Greek words. These two ways can best be distinguished by different modern punctuation. Either one can say μανθάνει νῦν γράμματα, εἴπερ ἐμάνθανεν ἃ ἐπίσταται and mean "if one has learned what he knows, then he is now learning the alphabet," or one can say μανθάνει νῦν, γράμματα εἴπερ ἐμάνθανεν, ἃ ἐπίσταται and mean "if one has learned the alphabet, then he is now learning what he knows." As a single conditional statement admitting two readings depending on vocalization, this fits the pattern of C/D fallacies elsewhere in *S.E.* But in light of its similarity to the *Euthydemus* argument,[46] it seems more likely that the conditional signifies a premise and a conclusion to an argument, not a statement with an ambiguous transcription. To refute the claim that one learns what one does not yet know, the argument is exactly that of Euthydemus, with the sophistical assumption of a linguistic parts-whole identity:

> One knows (= has learned: ἐμάνθανεν) the alphabet.
>
> [What one is now learning is composed entirely of the alphabet.]
>
> ———————
>
> Therefore, one is now learning that which one knows (ἃ ἐπίσταται).

The argument assumes the equivalence of what one has learned (ἐμάνθανεν) with what one knows (ἃ ἐπίσταται).

The emendation of Ross
Ross changes the imperfect verb in the premise to the present tense: μανθάνει νῦν γράμματα εἴπερ μανθάνει ἃ ἐπίσταται. Perhaps Ross's motive for keeping μανθάνει in both premise and conclusion is to retain the common term in both premise and conclusion. The argument might run:

> He learns (μανθάνει) what he knows.
>
> [He knows the alphabet.]
>
> ———————
>
> He is now learning (μανθάνει) the alphabet.

This reconstruction makes no use of the *Euthydemus* confusion, nor of any ambiguity in the transcription of the argument. At best, it seems to trade on some sort of equivocation on μανθάνειν, which would not be a case of Composition or Division. Moreover, to be persuasive the sense of the first premise must be that he *has learned* (ἐμάνθανεν) what he presently knows, which returns us to the original reading. For these reasons, I find this emendation and interpretation unlikely.

The emendation of Pickard-Cambridge
Pickard-Cambridge reads μανθάνει νῦν γράμματα εἴπερ ἐμάνθανεν.[47] By omitting ἃ ἐπίσταται, he translates the line: "He knows now if he has learnt his letters." The suggestion, here made very elliptically, is that knowing letters is sufficient for the knowledge of anything else. This is, of course, the point of the *Euthydemus* sophism, and I do not see how the omission of ἃ ἐπίσταται makes the argument any clearer.

The emendation of Poste
Poste offers a particularly elaborate (and for that reason, doubtful) emendation: καὶ μανθάνων γράμματα ἅπερ μανθάνει ἐπίσταται: "by learning

the alphabet he knows what he learns." Once again, the attempt here seems to be to arrive at something that mirrors the argument of the *Euthydemus*. Such an attempt seems to me to be correct, but if that same argument can be sufficiently captured without emendation, I would prefer to stay with the received text.

We are left with, it seems to me, two options. We might interpret the example as Forster does, as another illustration of a Greek transcription that permits two different vocalizations. In that case, this is an example of Composition entirely at one with the *S.E.* examples discussed earlier in "Further Examples." On the other hand, we might read the example as an abbreviated argument on the model of the *Euthydemus* argument. The sophistical error of Composition under this interpretation lies not in the various ways the conditional might be pronounced but in the suppressed premise that equates the alphabetic (or phonetic) στοιχεῖα with wholes composed from those letters (or phonemes). Aristotle may still call such a confusion one of C/D, for it arises from an inability to distinguish different signifiers composed of the same elements. It only differs from the usual examples of C/D by not admitting an oral disambiguation.

That same sophism also may be at work in another example that has vexed commentators. In his *S.E.* 20 list of examples of errors due to Composition, Aristotle refers to

> the argument of Euthydemus: Don't you know now, being in Sicily, that there are triremes in the Piraeus?[48]

Unfortunately, we cannot look to Plato's *Euthydemus* for help in reconstructing this argument. None of the sophisms recounted there has any similarity to this example. On the surface it might appear to be an argument similar to the "being born now" argument, that is, one that plays on the location of the temporal modifier.[49] But the mere presence of νῦν should not mislead us.[50] What complicates such a line of interpretation—indeed, what complicates all lines of interpretation—is Aristotle's reference in the *Rhetoric* to the same argument, again as an illustration of C/D.

> Such is the argument of Euthydemus, to know that there is a trireme in the Piraeus, for he knows each one.[51]

In this brief mention of the argument, there is no temporal reference, nor reference to anyone or anything being in Sicily. Interpreters of this line have sought to identify two things (ἕκαστον), each of which is known separately but not known in conjunction. The most popular reconstruction of this example is that of Schrader, repeated by Cope and endorsed by Grimaldi, namely, that "What you know, you know in the Piraeus where you now are.

And you know that there are triremes (i.e., that they exist somewhere). Therefore you know that there are triremes in the Piraeus (even if there are not)."[52] Unfortunately, as Cope notes, such interpretations conveniently ignore the *S.E.* 20 presence of ἐν Σικελίᾳ ὤν. The false conclusion that seems to be wanted in *S.E.* is that someone who is *not* in the Piraeus nevertheless has knowledge of what is in the Piraeus, rather than that someone *in* the Piraeus has knowledge of something being there which is not there. I can offer no resolution to the problem of reconstructing this fallacy, except to note that the *Rhetoric* explanation, ἕκαστον γὰρ οἶδεν, need not refer to only two separate items. (One might have expected ἑκάτερον if it did.) Freese translates "because he knows the existence of two things," which, on any account, sounds overly metaphysical for this linguistic fallacy. Cope renders it "because he knows each of two things which are here omitted."[53] A better reading is that of Roberts (Revised Oxford): "because he knows the separate details that make up this statement." The level of στοιχεῖα in question may be two propositions, but it also may refer to the separate words, or even the component γράμματα of the final conclusion. Admittedly, the sophism loses much of its interest if it is reduced to another version of the stale equation of a proposition with its component words or letters. But reading the *Euthydemus* teaches us that no fallacy is too puerile not to be allowed almost tedious repetition in sophistic circles.

In summary, we may say that Aristotle's designation of C/D fallacies as not being examples of double meaning entails that they must involve an inability to properly individuate between two different verbal signifiers (not one verbal signifier with two meanings), and his further insistence that C/D fallacies are "due to language" entails that they do not involve confusions of extralinguistic parts and wholes. Nevertheless, the example from *S.E.* 4 ("knowing letters") and the example from *S.E.* 20 ("triremes in the Piraeus") may suggest that some confusions between parts and wholes are included among C/D fallacies, namely, confusing a linguistic signifier with the more basic linguistic components that make up that signifier.

C/D Fallacies in the *Rhetoric*

The "knowing letters" fallacy and the "triremes in the Piraeus" fallacy mentioned in *S.E.* are two of the five illustrations of fallacies due to C/D offered in *Rhetoric* II. To complete our study, we should look at the three other *Rhetoric* examples, two of which are not easy to accommodate within the *S.E.* analysis.

The first example is fairly close to several *S.E.* illustrations.

Example 1 And since twice so much makes one ill it is false to claim that a single portion can make one healthy. For

> it makes no sense that two good things are one bad thing. In that way then it is able to refute, but in this way it is able to prove: for one good thing is not two bad things.[54]

The best way to interpret this fallacy along the lines of the *S.E.* examples is by analogy with "five is three and two":

> If this single dose produces health, and that single dose produces health, then this-and-that-dose (i.e., two doses) must produce health.

> Example 2 Again, what Polycrates said in praising Thrasybulus, that he overthrew thirty tyrants, for he combines them.[55]

Polycrates had argued that Thrasybulus deserved thirty rewards for his part in overthrowing the Thirty Tyrants. Aristotle regards this as an example of argument by fallacious Composition, because the Thirty Tyrants made up one tyranny, not a combined total of thirty tyrannies. This example does not seem amenable to the *S.E.* analysis of the C/D fallacy type. It involves a conceptual confusion, abetted by expressions of ordinary language. Although it is common for many members to compose a single distinct class, it is rare that one and the same class can be composed of only one of its members. So, for instance, we deny that a single soldier can make up an army or that a single senator can make up a senate, but we allow that a single tyrant can make up a tyranny. But more properly, tyrants do not "make up" tyrannies; they bring about, or cause, tyrannies. The problem is that a tyranny is not a class at all but a condition whose sufficient cause is a single tyrant. Strictly speaking, it mistakes a condition for a class and turns sufficient causes into class members. The error lies at the conceptual level, though that conceptual error is encouraged by such linguistic usages as "the tyranny was made up of the Thirty." In this respect, the error may partially derive from ordinary language, but it rests upon a conceptual confusion between classes and conditions.

The final example from the *Rhetoric* is specified as a case of Division:

> Example 3 And in the *Orestes* of Theodectus there is a fallacy from Division. If she should kill her husband, it is just that she should die, and [it is just] for the son to avenge his father, and so these things were done [by Orestes]. For when combined perhaps they are no longer just.[56]

The argument seems to be in reference to Orestes whereby (1) the justice of a husband-slayer being put to death and (2) the justice of a son avenging the death of his father leads to the false conclusion that (3) it was just for Orestes to slay his mother. We might put it this way. If X slays Y, then Greek justice demands that:

1. X should be slain for killing Y (where Y is a father), and
2. Y should be avenged by Z (where Z is Y's son).

   The (fallacious) practical conclusion drawn from 1 and 2 is that it is just that

3. Z should avenge Y by slaying X,

   for it fulfills both prior claims of justice in a single act. But if the act of Z slaying X is one of matricide, then it turns out to be unjust. Such is the moral dilemma that Orestes is placed in by the prevailing norms of Greek justice.

It should be noted that this is not strictly an argument from the justice of 1 and the justice of 2 to the justice of (1 and 2). Action 3 is not the same as the conjunction (1 and 2). Actions 1 and 2 are two general types, whereas 3 is a different action at a higher level of specificity. The fallacy is a result of mistaking two acts of different specificities to be one and the same act. The resolution would require a distinction between the two acts. This, like all fallacies of Composition and Division, is not a case of double meaning. There is no single account that signifies two acts. Rather, it is a failure to distinguish between two accounts, each signifying a distinct act. Just as in the case of the "Thirty Tyrants" argument, ordinary speech may encourage the belief that being slain and being slain by Z are identical acts. But unlike the *S.E.* examples of Division and Composition, pronunciation or vocalization cannot disambiguate between the two signifiers. Aristotle is consistent in noticing that, unlike fallacies of double meaning, here it is the failure to distinguish signifiers that leads to a failure to distinguish between acts signified. But the cause of that initial failure seems to have little to do with a distinction between utterances and their transcriptions.

In summary, not all of the *Rhetoric* examples of C/D fit neatly into the more developed *S.E.* scheme. Such difficulties, however, need not discourage scholars from studying Aristotle's *S.E.* project on its own merits.

CONCLUSION

In this chapter I have dealt with several obstacles to understanding Aristotle's inclusion of fallacies due to Composition, Division, and Accent as "due to language" but not "due to double meaning." First, there is a long tradition of interpreting fallacies due to Composition and Division as parts-whole confusions. Most parts-whole confusions, however, are not due to language. The only Aristotelian warrant for such an interpretation derives from a mistranslation of a text from *Rhetoric* II, 24, and a couple of problematic examples from that chapter. I suggest that one should circumvent this obstacle by not expecting to harmonize completely the *Rhetoric* treatment of fallacious enthymemes with the *S.E.* treatment of fallacious syllogisms. I have reserved most of my discussion of the *Rhetoric* examples to "Confusing Linguistic Parts and Wholes" and "C/D Fallacies in the *Rhetoric*." The major exception to my policy of avoiding the *Rhetoric* material is to accept Aristotle's treatment there of Composition and Division as a single type of fallacy. I present reasons for believing that Aristotle is committed to this position even in *S.E.* (despite his taxonomical distinction) in part 2 on Resolutions.

The second matter of concern was why fallacies due to C/D are not cases of double meaning. The key to answering this question lay in Aristotle's repeated association of C/D fallacies with those due to Accent. The latter were shown to be chiefly written fallacies. The distinction between how one may differently utter a Greek sentence and how such different utterances are identically transcribed in fourth-century B.C. Greek explains why mistakes due to C/D appear to be but are not really amphibolous double meanings. The key to understanding Aristotle's position is his belief that different linguistic signifiers are individuated by how one vocalizes a sentence, not by the sequence of component parts of the sentence. I have illustrated that principle by reviewing Aristotle's examples of C/D throughout *S.E.*

Finally, I note that my account of Aristotle's fallacies of Composition and Division also can be found in William of Sherwood's thirteenth-century *Introductiones in logicam*. William writes:

> Composition is the connection of things that are more disposed to be compounded. Division, on the other hand, is the separation of things that are more disposed to be divided. I am speaking here of connection and separation in the act of speaking (in actu proferendi).[57]

Later, William adds:

> "Composition" indicates one act of discourse and "Division" another, both acts, to be sure, being based on a single substance of utterance.[58]

This emphasis on the utterance rather than the written proposition is the major means by which Aristotle distinguishes between errors due to Accent, Composition, and Division and those due to double meaning (homonymy, amphiboly, and Form of the Expression). In the former cases, one fails to recognize that a unique signifier is being used with multiple significations. In the later cases, one fails to recognize that two unique signifiers, each with univocal signification, are being treated as one signifier. In the one, the appearance of syllogistic validity derives from ignorance of multivocity. In the other, the appearance of syllogistic validity derives from ignorance of what individuates a signifier. These different sources of the false appearance are Aristotle's chief means of distinguishing between fallacy types. Isolating these sources is the task of resolving the fallacy. I turn now to Aristotle's crucial notion of the resolution of false reasoning.

Part 2

Resolutions of False Arguments

Chapter 5

Resolutions of False Arguments

Introduction

There is a natural division in *S.E.* between chapters 15 and 16. In the first half of the treatise, Aristotle is mostly concerned with the production of apparent arguments. The question being addressed is "how does the sophist accomplish the effective simulation of argument?" The answer is that he exploits the six linguistic and six nonlinguistic sources of false appearances. In chapter 16, however, Aristotle introduces a new concern: the problem of false reasoning from the standpoint of the potential victim of the sophism rather than from the standpoint of the perpetrator. His concern is with resolutions (λύσεις) of sophistical arguments. The organization of his material on resolutions parallels his earlier format. He devotes chapters to each of the types of fallacies, both linguistic and nonlinguistic, and shows via examples and commentary how each type is to be resolved. There is, however, in this new material an odd feature. One would expect that the resolution of a false argument would only require the exposing of some particular flaw in the argument, but Aristotle recognizes that fallacious arguments may have multiple flaws. Exposing just any flaw does not constitute a resolution: one must expose the right flaw.[1] One task of this chapter is the investigation of the principles that Aristotle employs to isolate that single characteristic error of any example of false reasoning.

An important by-product of this investigation will be a better appreciation of the rationale underlying Aristotle's overall typology of fallacies. As a result, two difficulties already noted in Aristotle's treatment of the six fallacies due to language will be alleviated. Recall that Aristotle has divided these six types of linguistically based fallacies into two subgroups: those due to double meaning and those not due to double meaning. Within each subgroup, Aristotle tries to show (1) significant differences among the three members to justify the threefold distinction, and (2) significant similarities among the three members to justify their unity as a subgroup truly distinct from the other subgroup. First, concerning the differences among the members of the subgroup not due to double meaning, it was noted that the fallacies of Composition and Division do not seem to be two different kinds of error. They seem to be two different descriptions of the same mistake. Aristotle's doctrine of fallacy resolution helps explain why these two fallacies should not be distinguished. Second, concerning the members of the subgroup of linguistic fallacies due to double meaning, a difficulty was raised in chapter 3 about whether fallacies due to Form of the Expression really involve double meaning in the same sense as fallacies due to homonymy and amphiboly do. Again, Aristotle's comments on resolutions help clarify this issue, since proper resolutions turn out to be the distinguishing criteria in the overall typology of false reasoning.

PRINCIPLES OF ARISTOTELIAN ANALYTICAL METHOD

We need now to note two important principles of Aristotelian analytical method. These two principles help determine the specific architecture of his typology of fallacies. First, one can only understand defective examples of any process or species if one first understands what it means to be a nondefective example. The study of normal or properly functioning types always must precede the understanding of abnormal or dysfunctional examples.[2] False reasoning is a defect of true reasoning, so any study of false reasoning must begin from an understanding of the characteristics of true reasoning.[3] Second, understanding something is, for Aristotle, knowing its causes, knowing how to explain it. Accordingly, to understand true reasoning is to know the causes of it. Reasoning is an activity, and the distinguishing explanations of activities are their moving (efficient) cause and, especially, their final cause. The moving cause of reasoning is wonder or puzzlement.[4] The final cause is the removal of that puzzlement, which translates into an understanding of why something is the case. False reasoning, simply because it is false, can never entirely relieve the mind's puzzlement or perplexity about the conclusion. Even if someone is entirely convinced by a piece of false reasoning and believes he now understands that which formerly perplexed him, Aristotle will only grant to him *apparent* relief from his initial puzzlement. Psychological relief is not a sufficient condition of intellectual relief. Here, as elsewhere in Aristotle's philosophy,[5] the distinction between a condition for us (ἡμῖν) and that condition by nature (φύσει or ἁπλῶς) is important. The felt relief from some mental difficulty is no guarantee of an actual end to that difficulty, so even the acceptance of some false argument fails to result in the loss of the perplexity that first moved one to undertake the process of explanatory reasoning.

Not only do false reasonings leave intact the original puzzlement that motivated the search for explanations, but when those false reasonings take the form of false refutations, there is an increase in confusion. A refutation is a species of reasoning that leads to the denial of some hitherto accepted claim. A false refutation only appears to lead to such a denial. Far from removing perplexity, the false refutation creates a puzzle where once there was none. Now one is faced with competing arguments, neither obviously defective, leading to contrary conclusions.[6] The previously held claim, while thrown into doubt, is not entirely overthrown, since one still has the original reasons for accepting it. To summarize, then, reasoning is undertaken to remove perplexity about some claim. If the reasoning is sound, the perplexity is dissipated. If the reasoning is false, the perplexity remains. If the reasoning is a false refutation, then a claim that previously engendered no perplexity, while not convincingly refuted, is thrown into doubt, creating additional perplexity.

How does this new perplexity affect the victim of the false refutation? Like all perplexity, it motivates the search for a resolution. And a true resolution is one that returns the victim to that state of mind prior to its disruption by the encounter with the false refutation. Resolutions, then, must be fully explanatory. They explain why the purported argument is a false argument. But given the defining character of false arguments, that is, that they appear to be what they are not, such an explanation actually must account for two different things: (1) It must explain why the purported reasoning is not real reasoning. This is a fairly objective procedure. Reasoning (συλλογισμός) is strictly defined, and there are a limited number of ways to violate that definition (see section below on *Ignoratio Elenchi*); and (2) It must explain why the false reasoning appears to be real reasoning. But such appearances are always appearances to someone. Someone has to have granted the premises and granted the inference to the conclusion.[7] Moreover, Aristotle recognizes variations in the abilities of people to detect false moves in reasoning. One and the same argument, he says, may be thought by some people to be deceptive because of a linguistic confusion and by others to be deceptive because of one of the nonlinguistic sources of fallacy.[8] The odd feature we noted above, namely, that it is not the exposure of just any flaw that constitutes a resolution, raises the following problem. Given that an argument may have multiple flaws (both linguistic and nonlinguistic), and that different people will assign the deceptive appearance of reasoning to those different sources, then if a proper resolution must explain why some false reasoning appears to someone to be true reasoning, it would follow that there could be more than one proper resolution to the same argument. Aristotle, however, regularly rejects alternatives to his prescribed resolutions to particular examples of false arguments.[9] How can he have it both ways? How can he criticize certain proposed resolutions while acknowledging that sophistical refutations are always relative to the abilities of individuals? In the next section I answer this question by appealing to a typically Aristotelian distinction between two kinds of resolution.

TWO KINDS OF RESOLUTION

Aristotle never explicitly answers the question raised above, but it is not hard to reconstruct the response that he would offer. He would invoke his distinction between what holds "relative to us" (ἡμῖν) and what holds "absolutely" or "without qualification" (ἁπλῶς), this time as it concerns resolutions. He already has paved the way in *S.E.* 8, where the distinction is applied to sophistical reasoning in general.

> A sophistical refutation is not an unqualified (ἁπλῶς) refutation, but rather relative to some person (πρός τινα). So too is [any sophistical] reasoning. For unless someone accepts that what is due to homonymy signifies one thing, and what is due to similarity of form signifies only a substance,[10] and so too with the other fallacies, there will be no refutation or reasoning, either unqualified or relative to the answerer. But if they should accept this, then there will be [reasoning] relative to the answerer, but not without qualification. For they [i.e., the questioners] have not secured [a premise having] one signification, but only the appearance of one and coming from that particular answerer.[11]

The reason sophistical refutations are always πρός τινα is that they work their confusion only if the victim holds some other false belief not explicitly part of the argument, for example, that some homonymous word is univocal, or that common nouns always signify particulars, or that certain nonsubstantial expressions always signify substance. Without such a false presupposition, there would not be even the appearance of a refutation. Resolutions, on the other hand, can be either relative to the person or relative to the argument.[12] When Aristotle prescribes a single resolution to a false argument, he is prescribing the resolution *relative to the argument*. When he canvasses other suggested resolutions, he rejects them on the grounds that they are only *relative to the person*. Both kinds, however, offer explanations of why the refutation is false and why it appeared true to the victim,[13] and the explanation of the appearance of truth always includes reference to some false presupposition held by the victim. The question, then, is how does Aristotle distinguish the false presupposition whose exposure leads to a resolution πρὸς τὸν λόγον from those other false beliefs whose exposures lead to the inferior resolutions πρὸς τὸν ἄνθρωπον?

One hint of an answer comes out of *Topics* VIII, 10, where Aristotle discusses arguments that derive a false conclusion, not through false reasoning *per se* but from false premises.

> One must resolve [such an argument] by destroying the false premise due to which the false conclusion arises.[14]

But what if the argument has more than one false premise? Modern logicians find it sufficient to show the falsehood of any premise in order to defeat the soundness of the argument. For us, any false premise qualifies as that "due to which the false conclusion arises." Not so in an Aristotelian resolution:

> For the one destroying any premise whatsoever has not resolved the argument, not even if the destroyed premise is false. For the argument may have more than one false premise.[15]

Aristotle believes that only one of the false premises is the real cause of the false conclusion, and that a proper resolution must determine which it is. He next offers two possible resolutions to an argument to a false conclusion. Suppose you know that Socrates is not (at present) writing. Aristotle considers the following apparent argument to the contrary:

	(1) He who sits is writing.	τὸν καθήμενον γράφειν.
	(2) Socrates is sitting.	Σωκράτη καθῆσθαι.
	————	————
Therefore,	(3) Socrates is writing.	Σωκράτη γράφειν.

Now you are thrown into new perplexity. Here is an argument that appears to refute a claim you know to be true. At this point I come to your rescue to resolve the apparent refutation and relieve your confusion about Socrates' writing status. Now let it also be the case that Socrates is not sitting. The wrong way to resolve the fallacy, says Aristotle, is for me to show you that the second premise is false. The reason is that even though my pointing out that Socrates is not sitting helps dispel your immediate perplexity about whether Socrates is writing, it will do nothing to dispel your perplexity should Socrates decide to sit down. I may have offered a (temporary) resolution relative to you, but it is not a resolution relative to the argument. For that, I must show you that the first premise is false, "for not everyone who sits is writing."[16] The one who points out the error in the first premise, says Aristotle, resolves the fallacy entirely (πάντως). That is, he accounts not just for this one token example of arguing to a falsehood, but for all related arguments of the same type, regardless of the particular second premise.

This characteristic of resolutions of arguments from false premises is applicable to resolutions of arguments with false reasoning. The difference, I suggest, between a resolution relative to the person and a resolution relative to the argument is the extent of the applicability of that resolution to other false arguments. To have resolved a false argument entirely (πάντως) is to have isolated that single erroneous belief that explains all fallacies of the same type. It might seem, then, that one must know the type of fallacy before one can determine how to resolve it. But Aristotle believes just the reverse: the type of fallacy is determined by the manner of resolution. We have, then, two claims. First, fallacies are classified according to their type of resolution. Next, the types of resolution are classified according to the extent of their applicability. In the following section, I investigate these claims.

THE PRINCIPLE OF PARSIMONY

The fact that Aristotle correlates fallacy types with resolutions appears throughout *S.E.*[17] We may state this correlation in two ways.

1. Each example of the same type of fallacy has the same resolution relative to the argument (πρὸς τὸν λόγον).

Because a resolution relative to the argument involves the isolating and exposing of just that false presupposition that makes the reasoning appear nonfallacious, we can restate the correlation as:

2. Each example of the same type of fallacy is resolved by isolating the same false presupposition.

The further claim that resolutions *determine* the type of fallacy becomes evident when Aristotle employs his principle, foreshadowed above in *Topics* VIII, 10, that the best explanation of false reasoning is the one with the widest applicability. I call this the Principle of Parsimony (PP) for reasons that will become evident. Aristotle employs PP to defend his classification of fallacies against competing classifications. We may state the principle in this way.

> PP: If one resolution schema S determines two classes of fallacy where another resolution schema T determines one class, then T is better than S.

In conjunction with the claim that fallacy types are determined by resolutions, PP is more than a tool for grouping particular examples of false arguments into more general types of fallacy: it defines what the different types of fallacy are. And by defining the types of fallacy according to the applicability of each proposed resolution, it also determines how many types there are. In short, PP picks out the defining characteristics of the *infimae species* of false reasoning. The final effect of such a project is an exhaustive classification of fallacy types.[18] Behind his twelvefold classification of fallacy types lies Aristotle's conviction that these represent the exact number of distinct sources of error, because it takes exactly these twelve types of resolution to explain them. And since resolutions of the same type of argument involve isolating the same false presupposition, the twelvefold taxonomy reflects Aristotle's bold conviction that there are exactly twelve types of false belief both necessary and sufficient to account for all occurrences of false reasoning.

PP can be invoked by Aristotle against three different sets of critics. First, there are those who believe that Aristotle underestimates the variety of ways that human reasoning errs. The vast and variegated gullibility in human rationality, say these critics, makes it impossible to produce a limited set of false presuppositions whose avoidance is sufficient to preserve one from all misreasoning.[19] Modern defenders of this position include De Morgan (1847): "There *is* no such thing as a classification of the ways on which men may arrive at an error: it is much to be doubted whether there ever *can be*"; Joseph (1916): "Truth may have its norms, but error is infinite in its aberrations, and they cannot be digested in any classification"; and Cohen and Nagel (1934): "It would be impossible to enumerate all the abuses of logical principles occurring in the diverse matters in which men are interested."[20]

Aristotle is no less aware of the great variety of human reasoning skills and temperaments than are these modern voices, yet such appreciation of human susceptibility to deception in no way lessens his confidence in the limited number of false presuppositions whose correction is sufficient to resolve all such errors. His response to these modern charges probably would be to point out that the "infinite" variety of error belongs to the human subject, not to the argument itself. The twelvefold taxonomy does not deny that people may be deceived in countless ways: it allows for innumerable resolutions πρὸς ἡμῖν. What it claims is that these potentially innumerable confusions can all be resolved by understanding a relatively limited number of facts about language and the extralinguistic world. My choice of the label

"Principle of Parsimony" derives from the use of the principle to limit the number of basic kinds of error that engender misreasoning.

Aristotle's second group of critics includes those who would claim that his twelvefold typology includes unnecessary redundancies. The most common form of this reductionist attack on the Aristotelian scheme comes from those who believe that all false reasoning is reducible to linguistic confusion.[21] Aristotle himself, in the opening remarks to *S.E.* on the "power of names," already acknowledged the inevitable possibility of linguistic confusion leading human reason astray. His six types of fallacy due to language reflect actual cases of linguistic deception, so Aristotle is quite cognizant of language as a fertile source of false reasoning. But to reduce all fallacy to linguistic confusion is tantamount to saying that being perfectly clear about the signification of words and phrases in an argument is sufficient to dispel all perplexity engendered by the fallacy. This Aristotle denies.[22] His method is to consider certain proposed linguistic resolutions to examples of fallacies outside of language and to show that although such resolutions may qualify relative to the person he can produce analogous examples of the same fallacy in which the language is indisputably clear and univocal. By PP these new fallacies must admit the same resolution as the original ones. As a result, the characteristic error of the originals, whose exposure explains away the perplexity caused by the false reasoning, must involve a false presupposition about the world, outside of language.

Aristotle's refusal to accept merely linguistic considerations as sufficient resolutions to false reasoning reflects a view of human reasoning in which the mind cannot rest satisfied—cannot be relieved of its perplexity (ἀπορία)—until it recognizes certain nonlinguistic facts about the world. It means that Aristotle, despite his appreciation of the linguistic bases of so much philosophical confusion, and despite his concern for investigating ordinary language distinctions for insights into the causes of such confusion, stands in stark contrast to, say, a Wittgenstein of the *Philosophical Investigations* on the therapeutic success of such concerns. In the only passages in which Aristotle does accept multiple effective resolutions encompassing both linguistic and nonlinguistic sources of error, the warrant for allowing both kinds of resolutions is that the linguistic resolutions *presuppose* the nonlinguistic.[23]

The final group of critics includes those who accept an exhaustive catalogue of fallacy types but believe that a number of different catalogues are equally acceptable. In the context of *S.E.*, these would be the people who offer alternative resolutions to some of Aristotle's examples without denying that his resolutions are equally legitimate. In the modern period, nineteenth-century logical systematizers such as John Stuart Mill and Richard Whately typify this approach.[24] To them, the exhaustive typology of false reasoning is pedagogically important, but the preference of one typology over another is purely a matter of pragmatic utility. Both Whately and Mill propose schemata of fallacious reasoning that are claimed to be exhaustive. Yet although exhaustive, neither system is offered with the stronger Aristotelian claims of objectivity and nonreducibility. Whately admits that "arbitrary choice" is often the only reason for placing some particular fallacy in one species rather than another, or even one species of fallacy under one genus rather than another. Mill concedes that actual examples of fallacies could be brought under more than one of his classes. These men, while appreciative of the Aristotelian project of codification, are not comfortable with Aristotle's confident claims that analyses of fallacies differing from his own are simply wrong.

Aristotle's distinction between resolutions relative to the person and resolutions relative to the argument goes some way toward explaining the alternative resolutions accepted by these critics. Still, his claim that resolutions relative to the argument result in a nonarbitrary and an objective taxonomy needs defense. Two procedures are open to him. First there is the inductive appeal: he might try to show the naturalness of his taxonomy by claiming as an empirical fact that exactly these twelve false beliefs prove to be both necessary and sufficient for the resolution of all examples of apparent reasoning. But even apart from this empirical claim, Aristotle also might appeal to an earlier-mentioned principle of his analytical method: the fact that false reasoning can only be understood by reference to proper reasoning. Remember, again, that resolutions must explain two things. They must account for why the reasoning fails to be true reasoning, and why it appeared to be true reasoning. The first explanandum presumes certain specific requirements necessary for proper reasoning. All that Aristotle needs to show for this half of the resolution is how the fallacy fails to meet one of those requirements. The specific number of requirements and the correspondingly specific number of ways to fail in those requirements make this part of the resolution quite nonarbitrary. In the next section, I present Aristotle's list of requirements for proper reasoning.

Proper Refutations and Their Defects: *Ignoratio Elenchi*

In *S.E.* 4 and 5, Aristotle mentions thirteen sources of fallacious reasoning. One of these, called "ignorance of what refutation is" (παρὰ τὴν τοῦ ἐλέγχου ἄγνοιαν, hereafter identified by its Latin label *Ignoratio Elenchi*), although introduced in *S.E.* 5 as a source of fallacy on a par with the other twelve sources, becomes in *S.E.* 6 an alternative way of incorporating all twelve of the other sources, since each of the twelve is represented as a violation of one or another part of the definition of a refutation. Aristotle defines a refutation in *S.E.* 5:

> Those fallacies due to the failure to define what reasoning is or what a refutation is arise due to a deficiency in the account [of reasoning or refutation]. For a refutation is a denial of one and the same [predication], not of a name but of a thing, and of a name which is not synonymous but the same name, and which follows necessarily from the premises set down (without including in the premises the original point to be demonstrated) and it denies it in the same respect and relative to the same thing and in the same way and at the same time.[25]

A real refutation, then, involves:

1. premises that do not include the conclusion;
2. a conclusion that follows necessarily from the premises; and
3. a conclusion that denies one and the same predicate affirmed by the answerer. That predicate denied must be (3a) the thing signified, not just the name; (3b) the thing signified by the same name affirmed by the answerer[26]; and (3c) the thing qualified in precisely the same way as it was affirmed by the answerer.

RESOLUTIONS OF FALLACIES DUE TO LANGUAGE

How These Fallacies Violate the Definition of a Refutation

The distinction between fallacies due to double meaning (homonymy, amphiboly, and Form of the Expression) and those not due to double meaning (Composition, Division, and Accent) is made clear when the particular clauses required of a true refutation are specified. Double-meaning fallacies violate clause (3a), above, whereas nondouble meaning fallacies *also* violate (3b) above.

It is easy to see how homonymy and amphiboly violate (3a). Take, for example, the claim that the same person cannot be both sick and healthy simultaneously. The sophist seeks to refute this with an argument purporting to predicate health of a sick person.[27]

> The person who became healthy is healthy.
> The sick person became healthy.
>
> ---
>
> The sick person is healthy.

This refutation, however, fails, because "the sick person" is homonymous. The conclusion truly predicates health of a person who *used to be* sick, but it fails to predicate health of a person who *is now* sick. But it is only the latter person whom the answerer claims cannot also be healthy. The victim of this false refutation errs in thinking that because "the sick person" is demonstrated to be healthy, that all persons possibly signified by "the sick person" are shown to be healthy.

Aristotle claims that false refutations due to Form of the Expression presuppose this same error on the part of the victim: the answerer falsely thinks that the fact (πρᾶγμα) that he stated is denied, and not just the name (ὄνομα).[28] In Category mistakes, the error is exemplified in this way. If one claims, say, that it is characteristic of an activity that one cannot simultaneously do x and have done x, the nominal denial of this claim is the admission that it is possible simultaneously to see x and to have seen x. It fails to constitute a denial of the fact, because seeing is not an activity but a "being-affected." The double meaning of "seeing x" is between "doing x" and "being affected by x." The difference between this kind of double meaning and fallacies based on homonymy and amphiboly is that in this case only one of the two significations is a correct signification. The name in question simply cannot have one of the two significations. In homonymy and amphiboly, on the other hand, the name or phrase can have either signification, although the one originally intended by the answerer is not the one made use of by the questioner. We can say, then, that homonymy and amphiboly involve confusions of two real significations of the same word or phrase, whereas Form of the Expression involves confusions of one real and one merely apparent signification of the same word. What unites these three types of error is that the first step of resolution, whereby the apparent refutation is shown not to be a real one, involves the same procedure: distinguishing between two meanings of a problematic word or phrase and showing that the meaning made use of by the questioner fails to refute the fact originally proposed by the answerer. All three errors violate one and the same requirement of true refutation.[29]

In linguistic fallacies not due to double meaning, the violation of (3a) above is compounded by the error of thinking that the name denied in the refutation is the same name as that affirmed by the answerer. In other words, fallacies of C/D and Accent violate (3b) as well as (3a). To take an example of a false refutation due to Accent, suppose one wishes to refute the (true) claim that a mountain always possesses spatial convexity. If one mistakenly believes that ορος is one and the same name regardless of pronunciation, then one might be deceived into thinking that because there exists a boundary (ὅρος) that is not convex, it is thereby shown that a mountain (ὄρος) need not be convex. As in the fallacies due to double meaning, (3a) is violated because convexity is denied only of the name ορος, not of each thing signified by that supposed name. This, however, is not a fallacy of double meaning, since ορος is *not* a single name with multiple significations. It is two different names, depending upon how one vocalizes it. As a result, not only are the things signified not the same (3a), but the names themselves are not the same (3b), despite the apparent identity of their written transcriptions.

Aristotle's own example of a violation of (3b) involves two names impossible to mistake as one and the same: "mantle" (λώπιον) and "cloak" (ἱμάτιον). The example, however, is intended to apply to the fallacies of C/D and Accent, where the two names are mistaken to be one and the same.

> Composition and Division and Accent are due to the phrase not being the same or the name being different. For this also would be required [i.e., 3b], just as it is required that the thing be the same [i.e., 3a], if one intends there to be refutation or reasoning. For example, if one intends to refute a conclusion about a mantle (λώπιον) one must not reason about a cloak (ἱμάτιον) but about a mantle. For while the latter conclusion [i.e., the one intended about the mantle] is also true, it has not been reached by reasoning, and there is a further need for a question whether it signifies the same thing in response to the one seeking the reason why there is a refutation.[30]

Aristotle's point is that just as λώπιον and ἱμάτιον are different names, so are, for example, ορος and ορος, depending on the vocalization.[31]

In summary, false refutations due to language are characterized by a denial that is only apparent (i.e., nominal) and not really of the thing conceded by the answerer. This occurs either when the same utterance only appears to signify the same thing (cases of double meaning), or when what only appears to be the same utterance (cases of C/D and Accent) only appears to signify the same thing. The latter group incorporate two false appearances, reflected by the violations of both (3a) and (3b) of the definition.

The Unity of Composition and Division: *S.E.* 23

I have argued that Composition, Division, and Accent are distinguished from fallacies of double meaning because it is not the same word or phrase but only what appears (especially in writing) to be the same word or phrase that has double signification. But all six forms of fallacies due to language are alike in involving one (real or apparent) signification confused with another, so there is a sense in which all six forms of fallacy due to language involve (real or apparent) double meanings. In *S.E.* 23, Aristotle suggests a common method of distinguishing the two confused significations for resolution of the error. It involves an appeal to opposites, and it is offered as especially characteristic of resolutions due to language. More importantly, it confirms, I believe, his position in the *Rhetoric,* that Composition and Division are one fallacy. The appeal to opposites shows that Composition and Division are related as resolution-to-fallacy, not as two distinct types of fallacy.

> As a general rule, in [false] arguments due to language, the resolution will always be according to the opposite of that due to which there is the [false] argument. For example, if the argument is due to Composition, the resolution is by means of dividing, and if the argument is due to Division, the resolution is by means of combining. Again, if the argument is due to an acute (ὀξεῖα) accent, the resolution is a grave (βαρεῖα) accent, and if the argument is due to a grave accent, the resolution is an acute accent.[32]

In other words, there is no more reason to divide fallacies based on an ambiguously written sentence into those due to Composition and those due to Division than there is to divide fallacies based on an ambiguously written word into those due to Acute Accent and those due to Grave Accent.[33]

Aristotle's seeming inconsistency between regarding C/D as a single fallacy in the *Rhetoric* (and suggested here in *S.E.* 23) and as two distinct fallacy types elsewhere in *S.E.* can be accounted for by differences in the examples that he cites of such fallacies. In some examples, the two different vocalizations of a particular word sequence amount to a choice between combining a phrase or dividing a phrase. In such a case, it is natural to regard an argumentative move from one to the other as a fallacy due to Composition in one direction, or a fallacy due to Division in the other direction. Consider, for example, the particular argument of "sitting while walking."[34] If one argues from the claim that

(A1) "it is possible to sit | while walking"
to (A2) "it is possible to-sit-while-walking"

it would be natural to say that the error is one of Composition: combining what should be divided. On the other hand, if one moved from the claim that

(B1) "it is impossible to-sit-while-walking"
to (B2) "it is impossible to sit | while walking"

it would be natural to say that the error is one of Division: dividing what should be combined. In this example, the choice of vocalizations is between dividing or not dividing (i.e., combining) a phrase. If all examples of C/D were like this, Aristotle would have a clear distinction between the two ways of false reasoning. But more often than not, the choice of vocalizations is not between dividing and not dividing a phrase; it is between dividing a phrase in one place and dividing it in another place. In these examples, choosing to divide a phrase in one way necessarily involves choosing to combine it another way. The choice is not between *whether* to divide or not, but between *where* to divide, which is just the same as choosing *where* to combine.

Consider, for example, the fallacy of the good cobbler from *S.E* 20.[35] In order to refute the claim that a cobbler could not be both good and bad, the sophist argues that since a good person could be a bad cobbler, it follows that a good cobbler will be bad: ἔσται ἀγαθὸς σκυτεὺς μοχθηρός. His vocalization of the false conclusion is ἔσται ἀγαθὸς σκυτεὺς | μοχθηρός. The answerer resolves the refutation by pointing out that the true conclusion is ἔσται ἀγαθὸς | σκυτεὺς μοχθηρός, and that the true conclusion and the false conclusion, despite appearances, are not the same expressions (3b above) and do not signify the same things (3a above). But one cannot claim that the resolution any more involves combining what was divided than dividing what was combined. It necessarily involves both. The resolution requires a revocalizing of a sentence in whatever way it takes to point out the difference between the signifiers.

It turns out, then, that different examples either encourage or discourage thinking that a real distinction exists between Composition and Division. But since not all of the examples admit a clear distinction, it is best not to insist upon such a dichotomy. Perhaps such a realization led Aristotle to abandon the distinction in the *Rhetoric*. Regardless, already in *S.E.* 23 Aristotle's discussion of resolutions by opposites signals a recognition that Composition and Division stand related more as fallacy type to resolution than as two distinct fallacy types.[36]

The Extralinguistic Component of Resolutions to Linguistic Fallacies

We noted that resolutions πρὸς τὸν λόγον must explain why the purported refutation is not a real refutation and why it appeared to be so. The first task is accomplished by showing how one or more of the requirements for true refutation is lacking. For linguistic fallacies, this amounts to distinguishing the opposed significations (real or apparent) and showing that the purported refutation is only nominal. In cases of C/D and Accent, making clear the opposed significations is also sufficient for explaining why the false refutation appeared valid. Mistaking the word ὅρος for the word ὄρος, for instance, only involves a mistake about a linguistic signifier. The appearance of soundness has nothing to do with any mistaken beliefs about boundaries or mountains. The same is true of resolutions of fallacies due to C/D. The rectification of a false presupposition about the identity of linguistic components themselves is entirely sufficient to resolve those errors.

Such, however, is not the case with linguistic fallacies due to double meaning. Something more than a mistake about language requires correction if fallacies due to homonymy, amphiboly, and Form of the Expression are to be entirely resolved. The persuasiveness of these errors is built on a false presupposition about how particular signifiers relate to the things signified. And to properly understand a relationship, one must properly understand both of the *relata*. On the linguistic side, one must be clear about the multivocity of signifiers. On the extralinguistic side, one must be committed to definite metaphysical presuppositions about the world. These were already evident in fallacies due to Form of the Expression. We saw that one could only recognize the falsity of certain reasoning if one properly understood the different Categories of predication. To resolve such fallacies, then, two false beliefs must be corrected. The first is the belief that morphological similarities and syntactical similarities signify ontological similarities. To see the error of that, one also must be free from false assumptions about the ontological distinctions among the things signified by similar morphological and syntactical signifiers.

In cases of false reasoning from homonymy and amphiboly, there also is an ontological mistake that lends the argument an appearance of soundness. I discuss this extralinguistic component to these resolutions in chapter 9. There I will show that Aristotle argues that distinguishing between multiple significations of the same name or phrase amounts to distinguishing between different states of affairs signified by an ambiguous premise. The importance of making that distinction is not just to determine which state of affairs is true. They may both (or all) be true. It is, rather, to be able to judge which premised state of affairs is part of the relevant explanation of the state signified by the conclusion.[37] In short, fallacies due to homonymy and amphiboly are finally resolved by an appeal to an explanatory structure among extralinguistic things and events. The sufficiency of linguistic clarification for explaining the appearance of refutation in false arguments turns out to be quite restricted; it applies only to fallacies of C/D and Accent where the identification and individuation of linguistic signifiers itself is misperceived.

The line dividing fallacies due to language and fallacies outside of language falls between those fallacies for whose resolution some sort of linguistic clarification is necessary, and those fallacies for whose resolution there is no need for linguistic clarification. In the remaining chapters, I show which ontological facts, besides the Categorial distinctions needed to avoid fallacies of Form of the Expression, Aristotle considers sufficient for resolving the remaining types of false reasoning.

Part 3

Fallacies Outside of Language

Begging the Question
Non-Cause As Cause
Accident
Consequent
Secundum Quid
Many Questions

Chapter 6

Begging the Question and Non-Cause As Cause

Introduction

A proper refutation, I have shown, requires the following conditions:

1. premises that do not include the conclusion;
2. a conclusion that follows necessarily from the premises; and
3. a conclusion that denies one and the same predicate affirmed by the answerer. That predicate denied must be (3a) the thing signified, not just the name; (3b) the thing signified by the same name affirmed by the answerer; and (3c) the thing qualified in precisely the same way as it was affirmed by the answerer.

Of the six fallacies outside of language, Begging the Question violates clause (1), *Secundum Quid* violates clause (3c), and the remaining four all violate clause (2) in various ways. I also have shown that a complete resolution of the confusion created by a false refutation requires more than identifying which clause of the definition of a refutation is being violated (i.e., why it is fallacious). It also must explain why the false refutation appeared not to violate that clause (i.e., why it was persuasive). A complete resolution, then, involves explaining away two perplexities that arise in tandem. When first confronted by an apparent refutation of something that one has believed true, one experiences perplexity over that conclusion. Is it true or not? One has had reason to believe it so, but now here is an argument appearing to refute it. By identifying the definitional requirement that the intended refutation fails to satisfy, that perplexity is resolved. The challenge to the original belief is shown to have failed. But a second perplexity arises. Why did the failed refutation seem so convincing? If the first perplexity was logical, this second aporia is epistemological: why was it that someone was (almost) persuaded by the false refutation? Resolving this perplexity involves, for Aristotle, the identification of certain underlying false presuppositions. When these presuppositions deal solely with ontological matters, Aristotle classifies such fallacy types as "outside of language." The aim of this chapter is to isolate those facts about the world, facts independent of language, about which clarity is necessary to preserve one from being persuaded by fallacies of Begging the Question and Non-Cause As Cause. I treat these two fallacy types together, because both derive their persuasiveness from false presuppositions about the proper explanatory relations among (nonlinguistic) facts.

THE FALLACY OF BEGGING THE QUESTION

The fallacy of Begging the Question[1] is one of the clearest examples of the importance of the nonlogical (and nonlinguistic) component required of some Aristotelian resolutions. To appreciate this, however, it is necessary to look at the treatment of this fallacy in the *Prior Analytics* and the *Topics*, for the fallacy is only briefly referred to in *S.E.* Presumably Aristotle believed that his detailed discussion at the end of the *Topics* (VIII, 13) required of him only a cursory discussion of this error in the immediately following treatment of fallacies in *S.E.* In each of his treatments, Aristotle establishes the error as epistemic rather than purely logical. And behind that epistemology lies a particular ontological order of nature.

Begging the Question in the *Prior Analytics*

Aristotle introduces Begging the Question in *Pr. An.* II, 16, as a failure of demonstration by distinguishing it from several other ways in which demonstration goes wrong. Demonstration (ἀποδείξις) is a sub-class of reasoning (συλλογισμός). Although a failure in reasoning invalidates a demonstration, there are more requirements relating premises to a conclusion in a demonstration than those specified in the clauses of Aristotle's *S.E.* definition of reasoning. These additional requirements are epistemic rather than logical. They guarantee an advance in someone's understanding rather than just logical entailment.[2] Two of the epistemic requirements placed upon premises in scientific demonstrations are that they must be better known (γνωριμωτέρων) than the conclusion and prior (προτέρων) to the conclusion.[3] These relationships can be understood in two ways: better known (or prior) to us (πρὸς ἡμᾶς or ἡμῖν), or better known (or prior) by nature (τῇ φύσει or ἁπλῶς). A proper demonstration must have premises better known and prior by nature, whether or not they also are better known and prior to us. One of the prescientific goals of dialectic is to bring it about that things that are better known or prior by nature become better known and prior to us.[4] The mental training of dialectic sensitizes the inquirer to this natural epistemic order. It leads him back, not just to first principles but to the recognition of them as first principles.[5] If the premises of a scientific demonstration are not better known than and prior to the conclusion, the demonstration fails, for the premises have not done their epistemic work of promoting understanding. But these failures, says Aristotle in *Pr. An.* II, 16, are not due to Begging the Question.

The failure of Begging the Question depends upon another distinction among propositions:[6] those that are known through themselves (δι' αὑτῶν), and those that are known through other things (δι' ἄλλων). The former are the first principles for which no other epistemic justification is possible.[7] The latter are not self-explanatory but admit justification by appeal to more evident claims (i.e., claims better known and prior). Aristotle accounts for the fallacy as a failure to recognize the proper members of these two classes of things or events signified.

> When someone tries to show through itself what is not known through itself, then he begs the question.[8]

Aristotle sketches two ways that this might happen.

1. Directly, one might appeal simply to the self-explanatory nature of the claim: "p because p," when p is not self-explanatory.
2. Indirectly, one might justify a claim through other claims that themselves require justification by appeal to the original point: "p because q, q because r, r because p." Although this looks like a recognition that p is not self-justifying, it reduces to "p because p."

It hardly needs pointing out that such circular arguments are logically unassailable. The importance of the *Prior Analytics* introduction to the fallacy is that it places the error in a thoroughly epistemic context. For Aristotle, some reasoning of the form "p because p" is acceptable, namely, in cases where p is self-justifying.[9] In other cases the same (logical) reasoning commits the error of Begging the Question. Distinguishing self-evident from non-self-evident claims is a notorious crux in the history of philosophy. Aristotle's antidote to the subjectivism that threatens always to debilitate such decisions is his belief in a natural order of epistemic justification and the recognition that it takes special (dialectical) training to make that natural order also known to us.[10] In the context of the *Analytics*, such training is assumed to have been successfully undertaken. Given such an audience, Begging the Question directly ("p because p") is an uninteresting, because it is an unlikely, possibility. Aristotle dismisses it quickly in the *Prior Analytics* because, among qualified scientific demonstrators, discrepancies would not exist between what appears self-explanatory and what naturally is self-explanatory. The fallacy in these contexts arises when something that neither appears to be nor really is self-explanatory turns out to have been assumed unwittingly at the beginning of a series of epistemic inferences. Although psychological convincingness plays as much of a role as logical entailment in the task of demonstration,[11] for the master of dialectic these two always coexist. The demonstrator, a successful graduate from the contests of dialectical training, finds convincing only what should be convincing and only in the way that it should be convincing. His cognitive state is at one with the explanatory order in nature.[12]

Begging the Question in Dialectical Reasoning

When Aristotle shifts his analysis of reasoning from a scientific context to a dialectical context, the presumed coextension of what should be convincing and what is convincing no longer exists.

> Begging the question is, in demonstrations, a matter of the premisses being related in this way according to truth [i.e., according to the way things are], but in dialectical reasoning according to opinion [i.e., the way things appear to be].[13]

Among the dialecticians of *Topics* and *S.E.*, what appears epistemically convincing sometimes is not naturally convincing. False dialectical refutations thrive in the gap between what appears to us convincing (according to opinion) and what really is convincing (according to truth). This is not to say that the natural distinction between self-explanatory and non-self-explanatory facts or considerations of natural epistemological priority play no role in dialectical arguments. On the contrary, the examinational or peirastic function of dialectic aims to discover which seeming truths (ἔνδοξα) are real truths. As different endoxa undergo various dialectical tests of confirmation or disconfirmation, a natural epistemic order will emerge. To forbid the use of any scientific principles in a dialectical examination is to expect too strict of a separation between scientific and dialectical reasoning. Particularly where epistemic orders are agreed upon by the dialecticians, appeals to such orders are to be expected and encouraged.

Attempts to deny any common ground between scientific and dialectical arguments are, I believe, often motivated by a notion of dialectic as purely a competitive sport where the goal is to upstage one's opponent regardless of the truth of the claim under dispute.[14] That is one of the uses of dialectic, but it is not the chief function for which Aristotle writes the *Topics* and *S.E.* The difference is brought out in *Topics* VIII, 5:

> Among those engaged in competition, the goal of the questioner is to appear to produce [some affect] by all means, and that of the answerer not to appear to be affected. But in those dialectical meetings held not for the sake of competition but for the sake of examination and inquiry, there are as yet no rules defined for what the answerer should aim at, and what kind of things he should and should not grant in order to properly or improperly defend his claim. Since, then, we have nothing passed down to us from others, let us try to say something about it ourselves.[15]

This is the context in which Aristotle discusses the error of Begging the Question in dialectic. Concern for the *Prior Analytics* distinction between self-explanatory and non-self-explanatory claims is not lost. What is no longer assumed is that dialecticians will initially agree upon where that line is to be drawn. The shift from demonstration to dialectic renders the error more easily made; it does not change the fundamental nature of the error.

Aristotle describes five different ways in which the fallacy of Begging the Question can arise in dialectic. The first way corresponds to the direct procedure that Aristotle dispensed with so quickly in the *Prior Analytics* ("p because p"). The other ways present deceptive variations on the indirect procedure from the *Analytics*, where one explains a claim by what only appears to be a different and prior claim (or set of claims). Brief comments follow on each of the five ways. They illustrate how Aristotle tries to differentiate proper reasoning from mere logical validity by appeals to differing explanatory relationships among nonlinguistic facts.

1. The most obvious way for the error to arise is when the premise assumed is "the same" as the point to be argued. Such a blunder is not easy to miss when the identical formulation is used but is more deceptive "in cases of synonyms and when a name and a phrase signify the same thing."[16] Aristotle is thinking of synonymy in the modern sense (i.e., in the sense in which "mantle" and "cloak" are synonyms) rather than in the *Categories* sense.[17] It might seem that we have here an extralinguistic fallacy being explained by reference to a linguistic confusion. But the apparent linguistic confusion cited here—the failure to perceive when the same thing is signified by different names or phrases—is not one of the two general types of linguistically based fallacies already detailed in Aristotle's taxonomy. It is neither a case of double meaning, where different things are signified by the same name or phrase

(homonymy, amphiboly, Form of the Expression), nor is it a case of different things signified by different names or phrases misperceived to be the same name or phrase (Accent, Composition, Division). Aristotle regards the fact of words having multiple signification to be a linguistic phenomenon, but the fact that things have multiple signifiers he considers a phenomenon about the things. I might illustrate the distinction this way, where X1, X2, and X3 are extralinguistic things, and N1, N2, and N3 are names.

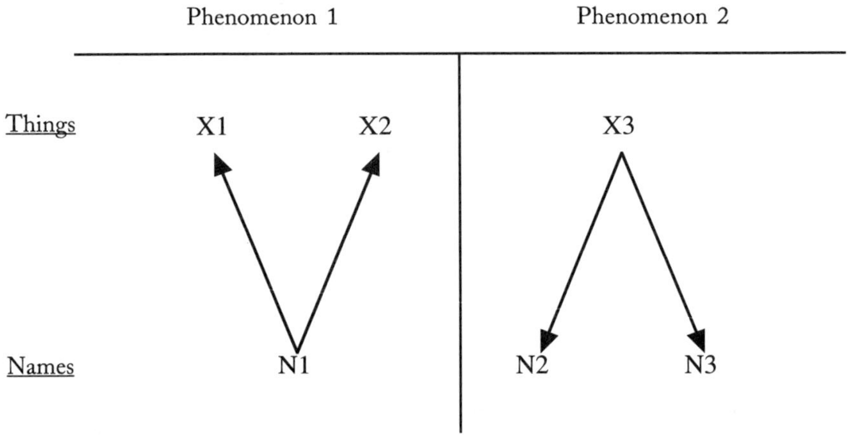

Aristotle considers column 1 to illustrate a phenomenon about language (specifically about N1). Column 2 illustrates a phenomenon about things (specifically about X3). What makes this distinction sound so alien to modern ears is that such a fact about a thing (that it is signified by different names) seems to carry no ontological significance. The fact that X2 has only one name and X3 has two names *by itself* seems to mark no real difference between the two things. It seems merely to reflect a conventional feature of a particular language. In short, while we must agree with Aristotle that the two phenomena are different (i.e., neither one entails the other), we would consider both to be features of language, not of the things signified. Aristotle, however, seems to regard some stronger distinction between 1 and 2 as reason for his classification of this confusion as extralinguistic. Directly Begging the Question because of variations in names or phrases is a case of misreasoning whose source is outside of language, because it reflects a fact about things, namely, that some are multiply signifiable, rather than a fact about names.[18]

2. The second way to Beg the Question is to try to demonstrate some particular predication by asking for the universal predication covering it. For instance, it is fallacious to try to prove that there is one science of contraries

(τά ἐναντία) by the concession that there is one science of opposites (τὰ ἀντικείμενα), since contraries are a particular example of opposites.[19] The problem here is entirely epistemic: the answerer cannot know the truth of the premise without prior knowledge of the truth of the conclusion.

3. The third way is difficult to interpret but seems to be intended as the converse of the second way. One cannot demonstrate a universal by assuming certain particulars under that universal. On the surface, this is a strange proscription, for it seems to render fallacious inductive arguments that generalize from particulars. This, however, cannot be Aristotle's meaning. In *Top.* I, 12, he cites induction (ἐπαγωγή) as one of the two approved types of dialectical argumentation, along with deduction (συλλογισμός). He further defines induction as

> a passage from particulars to a universal, for example the argument that if the expert navigator is the best [at his job] and the [expert] charioteer [is the best at his job], then too in general the expert is best at each particular job.[20]

He proceeds to praise induction as the more persuasive dialectical form of argument to an audience of ordinary people.[21] In short, inductive arguments are an integral, important part of dialectic. What, then, could Aristotle mean by this latter claim that one cannot demonstrate without fallacy a universal by assuming the particulars under it? The best way to make sense of this statement is to suppose that Aristotle is requiring that if one justifies a universal claim by appeals to particulars, those particulars must have some justification apart from the assumption of the universal. Admittedly, this interpretation seems to reduce this way of Begging the Question to the previous way (2) above. I see, however, no other interpretation that respects Aristotle's earlier commendations of inductive reasoning in dialectic.[22]

4. Another way of Begging the Question is to divide up the conclusion to be reached and to assume each of the parts. For example, if one wants to show that medicine is a science both of health and of sickness, one commits the fallacy if one assumes first that it is a science of health and later that it is a science of sickness. One cannot conclude "p and q" from the two premises "p" and "q" without Begging the Question.[23]

5. The fifth way to commit the fallacy is

> for someone to ask one or the other of two claims that follow from each other by necessity; for example, to ask that the side is incommensurable with the diagonal when one needs to show that the diagonal is incommensurable with the side.[24]

The error here is comparable to the first procedure, but with an important difference. Rather than justifying p by appeal to the semantic equivalence of p (i.e., p differently formulated), the fallacious appeal here is to the logical equivalence of p.[25]

Begging the Question and Immediate Inferences

This last case of fallacious reasoning has an interesting consequence. If an appeal to the logical equivalent of a conclusion as justification for that conclusion renders an argument fallacious, then certain deductively valid single-premised arguments (so-called immediate inferences) exemplify this form of misreasoning. An immediate inference is the entailment of a logically equivalent categorical sentence from another categorical sentence. Traditionally, such inferences have included the operations of sentence conversion, contraposition, and obversion.[26] Strictly speaking, Aristotle denies that immediate inferences constitute reasoning, because reasoning (συλλογισμός) requires at least two premises.[27] But more interesting for our purposes is that even apart from the formal defect of lacking a relevant premise, one cannot properly argue *from* a predication *to* its logical equivalent, because logically equivalent predications have the same natural epistemic force.

Logical equivalences may be epistemologically primitive (e.g., conversion equivalences) or epistemologically nonprimitive (e.g., contrapositives). In both cases, the component propositions of the equivalences have the same natural epistemic force. One cannot explain one claim by appeal to its logical equivalent. Nevertheless, the component members of nonprimitive equivalences may have different epistemic force to different people. To illustrate these two different situations (primitive equivalences and nonprimitive equivalences), I will consider Aristotle's dialectical treatments of claims involving term conversion and claims involving term contraposition.

Consider the relative epistemic strengths of a predication with its logically equivalent converse. Aristotle believes that the components of the two conversion equivalences ("No S's are P's" = "No P's are S's"; "Some S's are P's" = "Some P's are S's") possess the same epistemic force. Not only are equivalent converses epistemically equal, but in Aristotle's syllogistic theory those equivalences themselves are epistemically primitive. They are examples of self-explanatory truths: they are known through themselves. There is no possible argument available to convince the doubter of the equivalences. If one were to offer as epistemic justification for some claim its equivalent converse, one would be Begging the Question. To know one predication just *is* to know the converse.[28] What is sometimes referred to as an "argument" or a "proof"[29] that if no A's are B's then no B's are A's in *Pr. An.* I, 2 (25a14-17), is, strictly speaking, no such thing for Aristotle. Aristotle is producing dialectical reminders of the equivalence, not trying to justify it by appeal to a naturally prior epistemological principle.[30] It is dialectical because it serves to confirm that what appears obvious to someone really is obvious (because its denial leads to an immediately obvious contradiction).

Some logical equivalences, however, unlike converses, are not epistemological primitives. One member of a pair of nonprimitive, self-explanatory logical equivalents may appear more knowable and more convincing to a particular person at a particular time. This can be illustrated by comparing the case of a predication and its equivalent converse with the case of a predication and its equivalent contrapositive.[31] The primitive nature of the conversion equivalences makes it impossible, for example, both to believe that some philosophers are Greeks and to disbelieve that some Greeks are philosophers. But the nonprimitive nature of the contrapositive equivalences makes it possible, according to Aristotle, for someone, for example, both to believe that all philosophers are Greeks and to doubt that no barbarians (i.e., non-Greeks) are philosophers. In recognition of this possibility Aristotle encourages appeals to contrapositives as ways of defending or refuting a predication which, in its original form, seems indefensible or unassailable.[32] What he seems to mean is this. If one is trying to refute the claim, for instance, that whatever is noble is pleasant, and finds oneself unable to find an example of something noble that is unpleasant, he is advised to try to refute the contrapositive. That is, he should try to find something unpleasant that is noble. The search for something unpleasant among the set of noble things and the search for something noble among the set of unpleasant things are different psychological procedures, although aimed at the same discovery. The fact that Aristotle recommends that one consider both the predication and its contrapositive, when either defending or attacking it, shows that he recognizes that the two approaches are psychologically distinct yet logically equivalent. The refutation (or defense) of one of the pair may be more easily recognized (i.e., be better known) by person x, but when x recognizes that one of the pair is refuted, he is presumed to acknowledge that both propositions have been refuted. In short, Aristotle allows that the refutation of one of a pair of contrapositives may be more clearly recognized by a dialectician, but once he acknowledges *that* refutation, he also should recognize that the other member of the pair of contrapositives is likewise refuted.[33] It is the person who does not recognize the natural epistemic equivalences of logical equivalences who is vulnerable to a sophistical argument that Begs the Question. The remedy against falling into such confusion is greater dialectical training and discipline (ἀκριβέστερος), which will lead to a properly disposed mental state (τὸ τοῖς εὖ διακειμένοις τὴν διάνοιαν) where the illusion of a natural epistemic hierarchy among logical equivalences is dissipated.[34]

RESOLUTIONS

I have argued that Aristotle's fallacy of Begging the Question is not an error in logic. It arises out of a failure to recognize which states of affairs properly explain (and thereby justify beliefs in) other states of affairs. The failure is epistemic, but it is an epistemology grounded in the states of affairs themselves, not in the individual psychology of the believer. What appears to us to be known through itself may not naturally be known through itself, and things through which it appears to us that something is known may not be the things that naturally stand epistemically prior to that item of knowledge. There is an order of intelligibility built into the being of things, and the ignorance of that order by a dialectician renders him vulnerable to fallacies of Begging the Question.

Aristotle's advice in *S.E.* 27 for resolving fallacies of Begging the Question is brief. If one realizes that one is being asked to concede the original point, one should refuse to do so, even if the point being asked is a reputable belief. On the other hand, if one fails to realize that one has conceded the point at issue and the questioner uses the concession to produce the apparent refutation, then one should turn the tables on the sophistical opponent by oneself pointing out the fallacy committed. In dialectical exchange it is a worse mistake to be caught asking for the original point than to have inadvertently granted such a request. The answerer in such a position has failed to detect when different utterances mean the same thing.[35] The questioner, if he did not realize he was asking the original point, has committed the same error. But if he has knowingly asked for the original point, then he reveals himself to be ontologically confused: he has mistaken what is non-self-explanatory (known through other things) to be something self-explanatory (known through itself). In pointing this out to the false reasoner, one is not just pointing out a tactical psychological misjudgment by the questioner. It is not simply that the questioner falsely thought that the original point, if placed under the guise of a semantic equivalent, or a logical equivalent, or a covering universal, or divided up into exhaustive parts, would be more persuasive to the answerer. Rather, the questioner falsely thought that a non-self-explanatory fact about the world was an explanatory first principle. For Aristotle, that certain facts are self-explanatory while others are not is not a reflection solely of the cognitive abilities of humans. It is primarily a reflection of the structure of noncognitive reality. In short, a successful resolution of such a fallacy requires a firm grasp of the correct explanatory powers of things. Without a knowledge of which things are self-explanatory and which are not, the reasoner is liable to find a question-begging argument persuasive.

THE FALLACY OF TREATING A NON-CAUSE AS CAUSE

Arguments guilty of treating a Non-Cause As Cause (παρὰ τὸ μὴ αἴτιον ὡς αἴτιον) violate clause (2) of the definition of reasoning:

> for it is necessary that the conclusion follow by means of these things being so, which is not the case in fallacies of Non-Cause.[36]

Aristotle illustrates this fallacy by the example of an argument to the impossible (εἰς τὸ ἀδύνατον).[37] In such an argument, an impossibility is derived from a premise set that includes the claim targeted for refutation.[38] The error is to cite that premise as the "cause" of the impossibility when in fact that premise played no necessary role at all in the reasoning to the impossibility. What Aristotle means by calling premises "causes" (αἰτίαι) of the conclusion is that the state of the world signified by the premises explains the state of the world signified by the conclusion. They show why the conclusion has to be the case. And if one knows that it *has* to be the case, then perplexity cannot arise over its being the case. Premises are αἰτίαι insofar as they dissipate any perplexity over the truth of the conclusion. They generate εὐπορία by properly reflecting the nonlinguistic and nonpsychological explanatory relationships existing in nature.[39] What false refutations effect is the upsetting of that cognitive εὐπορία. They propose a counter-set of αἰτίαι, which only appears to remove any doubt as to the falsity of what hitherto has been confidently believed to be true. But far from removing perplexity, the false refutation creates a puzzle where once there was none.[40] To resolve such puzzlement the same two tasks already delineated are required. One must show why the apparent refutation fails to refute; then one must show why the failed refutation appears to refute. These distinct steps can be seen clearly if we now examine Aristotle's example of a Non-Cause fallacy and pay close attention to his instructions for resolution.

Suppose someone wishes to refute the common belief that the soul and life are the same thing. He begins with the assumption ultimately to be denied:

1. Soul and life are the same.

He then argues:

2. Coming-to-be is contrary to destruction.

Therefore

3. each particular kind of coming-to-be is contrary to some particular kind of destruction.
4. Death is a particular kind of destruction.
5. Death is contrary to life.

Therefore

6. life is a particular kind of coming-to-be, that is,
7. to live is to come-to-be.[41]

But (7) is impossible, therefore premise (1) must be false.[42]

Let me make two comments about this argument before looking at Aristotle's resolution. First, the absurdity of (7) may not be as evident to us as it would be to a Greek philosopher (or at least to a Greek philosopher trained in the Lyceum). For the Aristotelian, coming-to-be involved what we would call a substantial change: that is, the change from a previously nonexisting substance to a newly existing substance. "Living," on the other hand, was a process of change undergone by a substance that persisted throughout the change. So to say "to live is to come-to-be" would be immediately recognized as the absurdity of equating two mutually exclusive kinds of change. Second, I have reproduced the argument just as Aristotle relates it. There are, of course, certain suppressed premises necessary to make the inferences more strictly rigorous.[43] What we have, however, is enough for Aristotle to make his point.

In his resolution, Aristotle notes that the final absurdity (7) ("to live is to come-to-be") does not follow necessarily from *all* of the premises, since the same impossibility can be derived without (1). All one needs is (2) through (6) to reach (7).[44] The first premise turns out to be extraneous for the derivation of (7). This example shows that what Aristotle requires of reasoning by his stipulating "a conclusion that follows necessarily from the premises" is not only that the premises suffice for the conclusion, but that each premise is a necessary condition for the conclusion. In other words, there is a relevance requirement built into clause (2) of Aristotle's definition. False arguments "to the impossible" violate that relevance requirement. When Aristotle first introduces reasoning at *Top.* I, 1, this relevance requirement is explicitly included as a clause of the definition of συλλογισμός.

> Reasoning is an account in which, when certain things are set down, something different from the things that were set down follow by necessity *through the things set down.* (emphasis added).[45]

The emphasized phrase is what the error of Non-Cause As Cause violates, for the final falsehood is not derived through that particular premise set down for refutation.

What is important to note is that, for Aristotle, falsely believing that some premise is necessary for a conclusion is not simply to commit a logical error. It reveals a confusion about what sorts of things or events are properly explanatory of other things. The error is one of ontology. There is, for Aristotle, an explanatory structure built into things and events, independent of either the descriptions of those things or events, or the psychology of the particular reasoners. To adapt a phrase from Julius Moravcsik,[46] just as certain configurations in reality make a proposition true or false, so it is that certain configurations in reality make an argument either correct or incorrect reasoning. It is only because the universe exhibits these explanatory relationships that we are able to reason, whether properly or poorly. When a reasoner commits the fallacy of Non-Cause As Cause, he assumes a relationship between events or states of affairs that simply does not exist. These sorts of false arguments can appear to be true arguments through ignorance of the explanatory relationships that are part of the ontological structure of the universe.

It should be clear now why Hamblin cannot be correct when he states, "Since the workings of the Fallacy [of Non-Cause] can be analyzed in propositional logic it can be regarded as a formal one."[47] In the first place, Hamblin assumes that all fallacies due to Non-Cause have the form of false *reductio* arguments. But Aristotle never says this, and direct arguments with irrelevant premises perfectly match the general description that Aristotle gives us of this kind of misreasoning.

Second, under Hamblin's interpretation, the derivation of an impossibility from premise set {A, B, C} entails $\neg$ (A & B & C). The error, then, is to think that this in turn entails $\neg$A. But premise A may still be necessary for the derivation of the impossibility. The crucial move in Aristotle's resolution of the fallacy is to drop A from the argument altogether and see if the same impossibility still results:

> In those reasonings due to something additional being premissed, see whether when it is removed the impossibility follows nonetheless. After this [the answerer] must point this out, and say that he granted it not because it seemed true, but for the sake of the argument, yet the questioner made no use of it for the argument.[48]

It is clear that pointing out to the questioner that the impossibility persists even after the elimination of the targeted premise is enough to defeat the attempted refutation of that premise. Yet Aristotle goes on to ask something more of a complete resolution. Unfortunately, what that further requirement means is something of a puzzle. Aristotle seems to be offering the answerer a justification for initially accepting the superfluous premise. "Although I did not think it true, I conceded it because it seemed relevant to the argument. But you never used it in the argument." But why should the answerer have to justify his acceptance of the superfluous premise? The failure of the refutation has left the apparent truth of that premise intact. He need make no apology for a premise whose truth has in no way been rendered suspect.[49]

There seems to be no reason why Aristotle would require the answerer to offer some explanation for accepting the irrelevant premise, beyond the usual reason that he believes it to be true. There is, however, another premise in these false *reductio* arguments due to Non-Cause whose acceptance by the answerer does call for some special explanation, namely, the true cause of the impossibility. I suggest that, in this highly condensed text, when the answerer is instructed to "say that he granted it not because it seemed true, but for the sake of the argument," Aristotle is now referring to the premise that truly does cause the impossibility. Nowhere in his analysis of these false arguments does Aristotle suggest that the final impossibility has been invalidly derived. His claim is only that it has not been derived from the premise claimed by the questioner. In resolving the fallacy, the answerer must show that the impossibility still results from the other premises, all of which he has previously accepted. Thus far, the resolution removes the perplexity about the truth of the irrelevant premise. But at the same time it engenders perplexity about another of the premises. This new perplexity can be removed only by two further steps. The answerer must identify the premise that really causes the impossibility. Then he must explain why the real faulty premise was conceded in the first place. This last requirement is unique to those fallacies of Non-Cause in which the *reductio* form is used, for only in arguments "to the impossible" is there discovered a false premise previously conceded by the answerer. Nor is this last task merely an exercise in dialectical face-saving. It is required for the removal of all of the perplexity that necessarily arises when one concedes the validity of a *reductio* argument.

Aristotle does provide a warrant for granting a premise in dialectic that does not seem true to the answerer.

> The answerer should grant everything that seems true to him, and of the things that do not seem true, as many as are less improbable than the conclusion. For then he would seem to have argued sufficiently well.[50]

To say that a premise must be less ἄδοξον than the conclusion is to affirm the epistemic requirement of dialectic that we must argue from what is better known to us to what is less well known to us. To be shown that this subjectively "better-known" premise generates an impossibility is to be shown that what appeared more knowable to us is not more knowable by nature. This constitutes dialectical progress toward first principles and is the inevitable (and desirable) by-product of the resolution of a false refutation due to Non-Cause that employs the *reductio*. Two erroneous renderings of the explanatory configurations in nature have been rectified: that of the false reasoner, who claimed the wrong premise as explanatory of the impossibility, and that of the answerer, to whom the refuted premise seemed more likely than the original (false) refutation.

I have argued that three steps are necessary to fully resolve false refutations due to Non-Cause when a *reductio* argument is employed. First, show that the falsely refuted premise is not a cause of the impossibility. The procedure for this is to remove the premise and see if the impossibility remains (*S.E.* 29, 181a31–32). Second, identify the false premise that is the real cause of the impossibility, using the same procedure. Third, once the faulty premise is found, explain why that premise appeared acceptable when it was conceded. These last two steps are what Aristotle refers to at *S.E.* 29, 181a33-34. All three steps are demanded by Aristotle's overall theory of resolution as the alleviation of perplexity.

Returning to Aristotle's example of the false refutation that the soul and life are the same thing, can we complete steps two and three of the complete resolution? Already it has been shown that the falsehood "to live is to come-to-be" is generated without appealing to the identity of soul and life. The argument, then, does not place that belief in any jeopardy. But which of the remaining premises is the cause of the impossibility? Aristotle's answer would have to be (5): "Death is contrary to life."[51] For Aristotle distinguishes four different ways in which things can be opposed (ἀντικεῖσθαι): either as relatives (τὰ πρός τι), as contraries (τὰ ἐναντία), as privation and possession (στέρησις καὶ ἕξις), or as affirmation and negation (κατάφασις καὶ ἀπόφασις).[52] Whereas coming-to-be and destruction *are* proper contraries,[53] death is a privation of life, not a contrary to life.[54] What we have then is not false reasoning due to an equivocation on the sense of "contrary to" (ἐναντία). It is instead legitimate reasoning[55] leading from a false premise to an impossible conclusion.

Finally, why might the answerer have accepted the false premise, even though he may not have been convinced of its truth? Because it seemed more likely to be true than that the soul and life were not the same. It is easy to understand such a concession. After all, life and death are opposed, even if not as contraries.[56] It turns out that neither the claim that death is the contrary to life nor the claim that the soul and life are not the same are at all knowable by nature. Both are false. But it took the resolution of the false refutation from Non-Cause to make it clear to the answerer that what seemed more knowable to him was actually unknowable. The resolution, through all three of its steps, has not only resolved the two puzzles of why the refutation was false and why it appeared true, but it has advanced the real understanding of the answerer by exposing the error of thinking that death is opposed to life in some manner other than that of privation.

CONCLUSION

The two fallacies of Begging the Question and Non-Cause As Cause illuminate Aristotle's dichotomy between linguistic and extralinguistic sources of false reasoning. Neither fallacy results from any sort of linguistic confusion. The resolutions of both require the setting straight of errors about the proper explanatory powers of nonlinguistic facts. The possibility of Begging the Question arises from ignorance of which things are self-explanatory and which are not; Non-Cause As Cause arises from ignorance of which things are relevant explanations of non-self-explanatory facts.

Second, these fallacies shed light on the nature of reasoning (συλλογισμός) for Aristotle. Although both fallacies violate clauses of the formal definition of a refutation, neither amounts to a formal logical error. Reasoning, then, is more than just logical entailment: it requires a non-self-explanatory conclusion, all of whose premises are relevant explanatory factors.

Finally, these fallacies reflect the interplay of scientific principles with dialectical examination. If one of the aims of dialectic is to lead to scientific first principles, the dialecticians must already have some idea of the characteristics of a scientific first principle. Both fallacies can arise in either a scientific or a dialectical context. The difference seems to be that they will be found less often in scientific demonstrations, because proficient demonstrators are less prone to the ontological errors that engender the false reasoning.

Chapter 7

Accident and Consequent

Introduction

The fallacies of Accident (παρὰ τὸ συμβεβηκός) and Consequent (παρὰ τὸ ἑπόμενον) are naturally paired, Aristotle says, for the latter class of errors constitutes a special subset of the former class.[1] I argue below that Aristotle's attempt to distinguish Consequent from Accident fails. In actual fact, what we have here is one single fallacy with one single resolution.

Historically, there have been two common ways to analyze these fallacies. One is to account for them by appeal to linguistic ambiguity. This approach challenges Aristotle's fundamental distinction between fallacies due to language and fallacies outside of language, and it denies that examples of Accident and Consequent are outside of language. It is a challenge raised both by Aristotle's contemporaries, to whom Aristotle provides a response, and by many modern analysts, who have discussed the examples provided by Aristotle. The second approach treats these fallacies exclusively as errors in logical form. Such an analysis gained prominence in the fourteenth century under the influence of William of Ockham. In the modern period, this has been the prevailing analysis of the fallacy of Consequent. It was rechristened "the fallacy of Affirming the Consequent" by John Neville Keynes in the nineteenth century and continues to thrive under that label in most introductions to modern propositional logic. Whether the fallacies are deemed results of linguistic equivocation or of formal invalidity, absent from such analyses is any concern for the nature of the extralinguistic predications signified in the premises. Such a concern, however, is precisely what characterizes Aristotle's resolutions of these fallacies ἔξω τῆς λέξεως. In this chapter I examine Aristotle's examples of and commentary on these fallacies. I make clear the ontological appeals that he thinks are required for proper resolution. I also suggest several reasons why the non-Aristotelian interpretations of these fallacies as either linguistic confusions or formal errors became historically prominent.

FALLACIES DUE TO ACCIDENT AND THEIR RESOLUTIONS

Aristotle briefly introduces fallacies due to Accident (παρὰ τὸ συμβεβηκός) in *S.E.* 5. There he states how the fallacy is committed and explains why the procedure is fallacious.

> There are fallacies due to Accident whenever something is deemed to belong in the same way both to the thing (τῷ πράγματι) and to the accident. For since many accidents belong to the same [thing], it is not necessary that all the same [attributes] belong to all the predicates and to that of which they are predicated.[2]

We are dealing here with arguments of three terms. One premise predicates an accident of a subject (τῷ πράγματι). The other predicates a further attribute to either the subject or predicate of the first premise, while the conclusion predicates that same further attribute to the other term of the first premise. It should be possible, then, to exhibit these fallacies formally in the syllogistic style of the *Prior Analytics*. Thus what we might expect is an account of the fallacies in the manner of Aristotle's explanations of invalidity in the *Prior Analytics*: counterexamples to show the failure of the argument form to necessitate any conclusion. What we find though is something quite different. Aristotle ignores the form of the argument altogether and explains the error entirely by reference to the type of ontological predication found in one of the premises.

Aristotle illustrates the fallacy in *S.E.* 6:

> If a triangle has angles equal to two right angles and it happens that it is a figure or a first principle or a starting point, it does not follow necessarily that a figure or a starting point or a first principle has that predicate. For the demonstration is not *qua* figure, nor *qua* first principle, but *qua* triangle.[3]

The fallacious argument (with a brief schema next to it) runs like this.

All triangles have angles equal to two right angles.	T's are 2RA.
All triangles are figures.	T's are F.
————	————
All figures have angles equal to two right angles.	F's are 2RA.

The formal invalidity is evident, and it explains why the argument is false. But a resolution also must explain why the argument might appear true. To account for that Aristotle might have appealed to the formal structure of the second premise. He might have noted that only by converting the terms of that premise does the argument instantiate a valid argument form. Instead he appeals to the extralinguistic relationship between subject and predicate in that premise. It is that relationship which Aristotle calls "accidental." Figures (first principles, etc.) are only "accidentally" triangles.

But perhaps, one might say, Aristotle is not thereby making an ontological appeal. Perhaps he means by "accidental" only that the predication is non-necessary.[4] Whereas triangles are necessarily figures, figures are not necessarily triangles. Is not this just to make the formal point that the subject and predicate terms are not coextensive? And if the problem is simply one of mistaken extensions, there need be no further appeal to ontological kinds of predicates. In other words, by citing the "accidental" nature of the second premise, Aristotle need not be thought to be appealing to anything more than a failure of logical form. It becomes clear, however, when Aristotle discusses proper resolutions of these fallacies, that failure of coextension of predicates is not enough to explain the appearance of reasoning. More is needed if the resolution is to be complete.

The first step in resolving the fallacy is to show that it fails to meet all of the requirements for proper reasoning.

> Regarding those [fallacious arguments] due to Accident there is one and the same solution to all of them. For since it is undetermined when one should say [something] of a subject whenever it belongs to an accident (while sometimes it is thought to be so and people do say it, at other times they say that it is not necessary), it must be said then alike to all [such assertions] when the conclusion has been drawn, that it does not follow necessarily, and one must be able to bring forth an example.[5]

Aristotle notes that sometimes predicates belong to both subjects and accidents equally, and sometimes they do not. So it is not necessary that they do. The first step in resolving the fallacy is to show that the inference is not necessary (and so violates the definition of a refutation) by producing an example showing that such applications can fail.[6] This, however, is only a first step. Obtaining assurance of the failure of logical entailment is not the same as regaining assurance of the truth of the conclusion that was purportedly refuted. The reason for the appearance of refutation, as Aristotle says, is that sometimes predicates do belong equally to both a subject and its accidents. For restoration of complete mental confidence, the answerer needs to show not only (via the counterexample) that not all predicates belong equally to things and their accidents but that this particular predicate does not. Some have criticized Aristotle for not providing "a way to decide when the inference is allowed and when it is not."[7] That Aristotle needs some such decision procedure for complete resolution is correct. What these critics overlook is that Aristotle provides his answerer with such a test. It is not a test in the sense of some formal rule (such as convertibility) to apply to the terms of the accidental premise. It is, instead, an ontological test.

> For it is clear in all these cases that it is not necessary for what is true of the accident also to be true of the thing. For only to things that are indistinguishable and one in substance do all the same attributes seem to belong.[8]

This talk about things being "indistinguishable and one in substance" means more than things being coextensive or mutually necessary. It means that things must be related essentially. Furthermore, this relationship between subject and predicate of "belonging essentially" must be mutual. I understand this requirement to amount to the conjunction of the two types of essential (καθ' αὑτό) belonging distinguished in *Pst. An.* I, 4, 73a34–b5.[9] There Aristotle distinguishes between:

I. a predicate A belonging to B καθ' αὑτό because B cannot be defined apart from A (e.g., the way line belongs to triangle), and
II. a predicate A belonging to B καθ' αὑτό because A cannot be defined apart from B (e.g., the way odd belongs to number).

The fallacy of Accident arises in the triangle example when one interprets the second premise which is a καθ' αὑτό predication in sense (I) (since triangles could not be defined apart from figures) as if it were also a καθ' αὑτό predication in sense (II) (as if figures could not be defined apart from triangles).[10] The reason the sophistical argument about triangles might appear valid is that if the second premise were a statement of essential identity, the argument would be valid.

In summary, one resolves fallacies due to Accident first by showing, via a counterexample, that the premises do not entail the conclusion. This explains why the reasoning is false. Then one must explain that the deceptive appearance of true reasoning arises by thinking that one of the premises is an essential predication and show that the misunderstood premise is only an accidental predication.

False Resolutions to Fallacies Due to Accident

Next I consider a number of sophistical arguments whose deception Aristotle credits to Accident but which have received different accounts by other commentators. I show that it is the Principle of Parsimony (PP) in each case that prevents Aristotle from accepting the alternative analyses. Whenever an example seems to admit a purely linguistic resolution, Aristotle appeals to an analogous example that does not admit a linguistic resolution (e.g., the triangle argument above) and invokes PP to conclude that the nonlinguistic resolution that accounts for both examples must be the proper resolution. Aristotle's rejection of alternative resolutions depends upon his acceptance of PP. If that principle is rejected or ignored, many of the alternative accounts of these examples become defensible resolutions.

False Resolutions by Appeal to Linguistic Equivocation

Aristotle offers two false arguments due to Accident in *S.E.* 5. There are two alternative ways to try to account for them as due to linguistic equivocation. One is to cite the ambiguity of a word between its signifying a universal and a particular. We have already seen that Aristotle considers such confusions a common source of fallacies of linguistic double meaning.[11] The second way to analyze these arguments as linguistically based is to cite an ambiguity in the copula of one of the premises between the "is" of identity and the "is" of predication. Modern formalizations of these arguments in first order predicate logic with the identity relation lend themselves particularly well to this explanation.

The first argument reads as follows.

[A] Coriscus is different from [a] man.	ὁ Κορίσκος ἕτερον ἀνθρώπου
[Coriscus] is [a] man.	ἔστιν ἄνθρωπος
[Therefore,] he is different from himself.	αὐτὸς αὑτοῦ ἕτερος

If we understand the predicate "different from x" as "not x," and let c = Coriscus and M = man, the sophism is of the form:

c is not M.
c is M.
———
c is not c.

Any appearance of sound reasoning here seems to require a flagrant linguistic equivocation on the word "man" (ἄνθρωπος) between the two premises. The first premise is true only if "man" signifies the universal, but the second premise is true only if "man" signifies a particular man. Why then does Aristotle not consider such equivocation to be the source of the false appearance of reasoning? Because his second example cannot be resolved in the same way.

[B]	[Coriscus] is different from Socrates.	Σωκράτους ἕτερος
	Socrates is a man.	ὁ δὲ Σωκράτης ἄνθρωπος
	[Therefore, Coriscus] is different from [a] man.	ἕτερον ἀνθρώπου

Letting s = Socrates, we may schematize this sophism in this way:

c is not s.
s is M.
———
c is not M.

In this argument, there is no equivocation on ἄνθρωπος. In both premise and conclusion it signifies a particular man. By itself, the failure to account for [B] in the same manner as [A] need only show that the two arguments belong to different fallacy types. But Aristotle believes that there is a nonlinguistic resolution (*viz.,* Accident) that accounts for both, so the fact that he can offer the same resolution to both arguments means, by PP, that these are two examples of the same fallacy type. And because [B] is not a linguistic fallacy, neither can [A] be one.

One might object, however, that [B] really is an example of linguistic equivocation. One might say that it is demanding too much to require that the equivocation be on the same word for one to classify two arguments as being due to linguistic equivocation. Just as [A] is resolvable by pointing out the linguistic equivocation on "man," so too [B] is resolvable by citing a different equivocation. Bueno, for instance, writes:

> [Example [B]] may be regarded as a case of equivocation, for the expression "different from" has two different meanings. In the first premise it means "being a different individual from," and in the conclusion it connotes "not belonging to the (species) man."[12]

By this analysis, the argument equivocates on the sense of the negated copulas of premise and conclusion. Here Bueno seems to be alluding to the difference in modern logic between predication and identity. Let us see how the two arguments would be treated by modern logicians.

In argument [A]:

c is not M.
c is M.
―――
c is not c.

The first premise can be understood either as $c \neq M$ or $\neg Mc$.[13] The second premise, likewise, is either $c = M$ or Mc. For the argument to have true premises, it must be understood as:

$c \neq M$
Mc
―――
$c \neq c$.

The appearance of validity arises if the second premise is mistakenly thought to be $c = M$. It is this mistake that we commonly describe as confusing the "is" of identity with the "is" of predication. Argument [B] admits of a similar analysis. The sophism is intended to run as:

$c \neq s$
Ms
―――
$\neg Mc$.

In this case, however, even if Ms is mistakenly interpreted as $s = M$, the only valid conclusion would be $c \neq M$, not the desired $\neg Mc$. So there would need to be two different confusions of predication with identity to make this argument appear valid, one in the second premise and one in the conclusion.

This kind of analysis is the most frequent modern treatment of the pair of Coriscus sophisms.[14] Nor is it entirely removed from Aristotle's own treatment. His objection would be to our describing it as a linguistic confusion. For Aristotle, talk about confusing the "is" of predication with the "is" of identity masks an ontological difference between kinds of predicates. Aristotle never says of the grammatical copula "is" (τὸ εἶναι) that it is said in many ways. His claim is that what is (τὸ ὄν) is said in many ways. The copula itself, however, as a name has no signification. It only adds some signification, he says, to "some combination," that is, when it signifies an ontological relationship of belonging between two terms that do signify separately.[15] The danger,

I repeat, is not in Aristotle's formulation but in English renderings that sound as though the two readings of, for example, "Coriscus is a man," are due to some lack of clarity in the copula rather than due to multiple ways in which predicates can belong to subjects. Appeals to the identity relation of modern logic, he might say, only hide the ontological difference between accidental and essential predication.[16]

Even if one were to put aside the modern notations and concentrate simply on the double meanings of "man" in [A] and "different" in [B], Aristotle would still refuse to attribute the fallacies to linguistic error (i.e., to what Aristotle calls homonymy). What excludes them from the class of linguistically generated fallacies are such examples as the triangle argument of *S.E.* 6, which we discussed above. In that argument there is no possible appeal to verbal double meaning. The problem there is treating "All triangles are figures," as though being a triangle were not just accidentally related to being a figure (in that sense of "accidental," analyzed above). Aristotle assigns the same problem to [B]. The sophistical reasoner draws the paradoxical conclusion

> because the one whom he said Coriscus was different from [i. e., Socrates] just happened to be a man.[17]

The problem then lies in the accidental nature of the second premise.

To summarize, Aristotle presents us with three arguments:

T's are 2RA.	c is not M.	c is not s.
T's are F.	c is M.	s is M.
———	———	———
F's are 2RA.	c is not c.	c is not M.

Because Aristotle believes that he can explain all three by appealing to the accidental nature (i.e., the failure of essential identity) of the second premises, whereas there is no way of accounting for all three by appeal to linguistic equivocation, he concludes by PP that the error "due to which the false conclusion arises" is the ontological confusion, not the linguistic confusion (even if such linguistic confusions are what deceive some people). The strange statement that Socrates "just happened to be a man" must be understood analogously to "triangles just happen to be figures." The error is to think that all figures are essentially triangles, or that all men are essentially Socrates or Coriscus. In all three cases, the appearance of reasoning is credited to the fact that if the second premises were nonaccidental identity statements, the arguments would be nonfallacious.[18]

False Resolutions by Appeal to Oblique Context

In the next examples, Aristotle again attributes the deceptive appearance of reasoning to mistaking an accidental predication for an essential identity. What makes these examples interesting is that by Aristotle's own principles of resolution, they fail to exhibit the particular error that he posits.

> All these sorts of arguments are due to Accident. [1] Do you know what I am about to ask you? [2] Do you know the one approaching? or the veiled man? . . . For it is clear in all these that it is not necessary to truly affirm [the attribute] of the accident and of the subject.[19]

These two examples involve well-known identity paradoxes arising from failures of substitutivity in oblique contexts.[20] Aristotle illustrates the first argument by the following example:

> You do not know what I am about to ask you.
> What I am about to ask you is, e.g., what the good is.
> ———
> Therefore, you do not know what the good is.

The second argument runs:

> You do not know the one approaching/the veiled man.
> Coriscus is the one approaching/the veiled man.
> ———
> Therefore, you do not know Coriscus.

The form of these arguments is the same:

> You do not know A.
> A is B.
> ———
> Therefore, you do not know B.

This can be restated into more nearly Aristotelian (categorical) form as:

> A is unknown to you.
> A is B.
> ———
> Therefore, B is unknown to you.

Although either premise may be called accidental, Aristotle lays the blame for the fallacy on the second. The predications "what I am about to ask you is what the good is" and "Coriscus is the one approaching" are accidental. They are examples of things not being "indistinguishable and one in substance":

> In the case of the good, what it is to be good is not the same as what it is to be about to ask, nor is what it is to be approaching or veiled the same as what it is to be Coriscus.[21]

In the above text, the grammatical construction of the Greek articular infinitive εἶναι with a dative is Aristotle's standard way of expressing the essential natures of things. He nowhere denies that the logical form of the argument is invalid. Rather, he cites an ontological reason for that invalidity. The appearance of validity, he believes, is due to a mistake about the nature of the predication, not a misunderstood logical form.

Unfortunately, Aristotle underestimates the subtleties of arguments involving propositional attitudes. Even if A and B are essentially identical, the argument

You do not know A.
A is B.
―――――――
Therefore, you do not know B.

remains only apparent unless you also *know* that A is identical to B.

In contrast to Aristotle's (ultimately unsatisfactory) resolutions are those proposed by modern philosophers for these same difficulties. The two most prominent resolutions, Frege's (the error of confusing sense and reference) and Russell's (the error of confusing denoting phrases and proper names), both involve refinements in theories of reference not recognized by Aristotle. Neither requires any appeal to extralinguistic types of ontological predication. Of course, such Fregean or Russellian theories of reference, though effective for resolving certain paradoxes in oblique contexts, fail to account for the triangle and the two earlier Coriscus sophisms. Aristotle would conclude from this that, by PP, Frege and Russell have multiplied types of fallacy unnecessarily.[22] If Aristotle can show that the triangle argument and the oblique context arguments admit the same resolution, then he has shown that they exemplify the same type of fallacy. But we saw that Aristotle's own resolution of the oblique context arguments fails adequately to account for their appearance of validity. So by his own Principle of Parsimony, this failure of applicability suggests that Aristotle himself has illegitimately run together two types of fallacy: one type that arises in oblique contexts, which a refined theory of reference resolves, and a second type requiring a different resolution. Unless he can account for the fallacies of referential opacity under one of his other types of resolution, Aristotle's belief that his twelve fallacy types are jointly necessary and sufficient to explain all errors in reasoning cannot be maintained. Nevertheless, what I want to stress is that it is PP that Aristotle uses to adjudicate between competing typologies of false reasoning.

False Resolutions by Citing Missing Qualifications

Another of Aristotle's examples has baffled many commentators. All he says to identify the argument is, "Are small numbers multiplied by small numbers always small?"[23] Poste suggests an argument that is pure linguistic equivocation. As such, it is unlikely to be what Aristotle had in mind. His reconstruction looks like this:

> A small number is a small number.
>
> ---
>
> Therefore, a small number [even if] multiplied by a small number is a small number.

For Poste, "a small number multiplied by a small number" is ambiguous between referring to the first small number only or referring to the product (which could be a large number).[24]

A more promising reconstruction of the argument is offered by Joseph. He understands the false syllogism to be exemplified by the following:

> Six things are few (ὀλίγα).
> Thirty-six things are six things (namely, six distinct groups of six things).
>
> ---
>
> Therefore, thirty-six things are few (ὀλίγα).

Joseph explains the fallacy as lying in the second premise. "It is an accidental way of regarding 36 things, that they are six groups of six things."[25]

Joseph's example is an attractive reconstruction, but his analysis of the flaw is not entirely convincing. Arithmetical equations are problematic predications. Because numbers are conceived by Aristotle as sets of units, it would seem that the only predicate that could possibly belong καθ' αὑτό to a number would be the single unit itself. But not even the unit would belong to a numbered set in *both* senses of καθ' αὑτό belonging.[26] As a result, any arithmetic equation would be an accidental predication. But if this were so, one would expect the following argument to commit the same fallacy due to Accident.

> Six things are a number (of things).
> Thirty-six things are six things (namely, six groups of six things).
> ___
> Therefore, thirty-six things are a number (of things).

But this is a perfect first figure syllogism. When the predicate signified by "number" in the above valid syllogism is replaced by the predicate signified by "few," the syllogism is transformed, despite the appearance of formal validity. The reasoning is fallacious, because predicates such as ὀλίγον Aristotle assigns to the Category of relatives (πρός τι) and denies that such predicates can belong καθ' αὑτό to any particular thing.[27] Contrary to Joseph, then, the fallacy is engendered not by an accidental relationship between six squared and thirty-six, but by an accidental relationship of the predicate few or small (ὀλίγον) to six things (or to any other number of things).

Aristotle is particularly aware of how easy it is to misunderstand predications involving relatives. He singles out for criticism people who fail to acknowledge that arguments such as "the small number" involve any false reasoning at all. These people defend the reasoning on the grounds that thirty-six things *are* few (because fewer than, say, forty things).

> For if, when the conclusion has not been correctly derived, they pass over that fact and say that a true conclusion has been derived (for every number is both many and few), they are mistaken.[28]

This approach treats the argument as a failure to specify the particular thing relative to which the predicate truly holds. Other fallacies due to Accident were subject to similar alternative resolutions. One way people resolved the "approaching Coriscus" sophism was to grant the possibility of both knowing and not knowing the same thing.

> Some people resolve these by rejecting the question. For they say that it is possible to know and not to know the same thing, but not in the same respect. By not knowing the one approaching and knowing Coriscus, they say that they know and do not know the same thing but not in the same respect.[29]

These people granted the conclusion of the sophist but then added qualifications to show that the conclusion was not a true contradiction. They treated the "approaching Coriscus" argument as a linguistic fallacy of *Secundum Quid,* that is, of failing to specify qualifications.[30] Aristotle rejects this alternative analysis by again appealing to PP.[31] He cites two more examples of the same fallacy[32]:

That statue is yours.	That dog is yours
That statue is a work.	That dog is a father.
Therefore, that work is yours.	Therefore, that father is yours.
(= That statue is your work.)	(= That dog is your father.)

The subjects in the two cases are the statue and the dog. The accidents are a work and a father. The attribute of "being yours" is truly predicated of the subjects but not of the accidents. These arguments match the triangle example in form.

A is B.	That dog is yours.
A is C.	That dog is a father.
C is B.	That father is yours.

Aristotle offers two reasons for rejecting the *Secundum Quid* resolution to the approaching Coriscus argument. The first is that it fails to account for the paternal dog sophism. Even if it is true that one can know something in one respect and not know that thing in another respect, one cannot be someone's father in one respect but not be that person's father in another respect. What might work for "knowing x" fails to apply to "being x's father." The predicate "being a father" does not admit qualifications. PP, then, prevents the "approaching Coriscus" argument from being resolved as a fallacy of *Secundum Quid.*

The second reason he offers for rejecting the alternative resolution is opaque. Aristotle claims that, apart from the failure of *Secundum Quid* to account for the "paternal dog" example, the "approaching Coriscus" example itself fails to exhibit the necessary confusion among qualified senses of "knowing."

> Perhaps also in some cases nothing prevents this [resolution by *Secundum Quid*] from occurring, only this would not seem to be the case in the present example at any rate. For he knows both that Coriscus is Coriscus and that the one approaching is approaching. To know and not to know the same thing is thought to be possible, for instance to know that he is white but not to know that he is musical. For in that way he knows and does not know the same thing, yet not in the same respect. But as to the one approaching and Coriscus, he knows both that he is approaching and that he is Coriscus.[33]

I am unable to make Aristotle's meaning entirely clear. It seems that he is claiming that *Secundum Quid* fallacies involve knowing some distinct subject with respect to different accidents, whereas the "approaching Coriscus" fallacy does not involve two accidents distinct from the subject. Aristotle grants that it is possible to know of x that he is white and not to know of x that he is musical. This amounts to the claim that some subject P can be known to be Q and not known to be R. But in the "approaching Coriscus" fallacy, there are not three items of knowledge (P, Q, R) but only two. The sophistical conclusion is that the one approaching is known to be Coriscus and not known to be Coriscus. That is, P is known to be Q and not known to be Q. One might respond to Aristotle by saying that there are two different respects in which one may know and not know Coriscus. One can know that Coriscus is Coriscus but not know that Coriscus is the one approaching. Aristotle's answer seems to be that knowing Coriscus *qua* Coriscus does not count as some respect relative to which Coriscus is known. The *Secundum Quid* resolution requires two ways in which to know Coriscus, both of which differ from knowing Coriscus *qua* himself.

Final Remarks on Double Meaning and Fallacies Due to Accident

Aristotle's examples of fallacies due to Accident appear to us to be a variegated lot. With the lone exception of the triangle argument, all of them seem to admit an underlying linguistic confusion when analyzed by modern philosophers. In the two Coriscus arguments, there seems to be straight homonymy: "man" as signifying a particular or a universal and "is" as signifying predication or identity. In the oblique contexts, there is confusion between, for example, direct reference and indirect reference (i.e., sense). Or again, there may be a failure to specify qualifications, what Kirwan calls "equivocation by elision" and what Aristotle assigns to the fallacy type *Secundum Quid*.[34] Yet Aristotle's Principle of Parsimony effectively forestalls all attempts to resolve these fallacies by appeal to language alone. The Coriscus fallacies are not due to homonymy, because the triangle argument is not. Oblique context fallacies are not due to *Secundum Quid*, because the paternal dog argument is not. And the same principle prevents the paternal dog and statue arguments from being resolved as forms of linguistic equivocation. One might argue that in those arguments[35] "yours" (σός) signifies mere *possession* in the premises but something closer to *generation* in the conclusions. Works (ἔργα) and fathers are said to "belong to" people in a different and stronger way than pets and statues. But again, no appeal to linguistic double meaning can account for the triangle sophism, whereas the appeal to ontological confusion can account for the paternal dog sophism. "The dog is a father" is accidental, for being a dog is not essential to fatherhood, nor is being a father essential to doghood.

The crucial argument that unites these seeming disparate examples is the triangle argument. Because it is clearly not based on a linguistic confusion, and because that same extralinguistic explanation is able to account for all of the other examples, Aristotle classifies all of the fallacies as "outside of language." It is not the case that none of these examples can be understood as trading on verbal ambiguities; rather, it is the case that at least one of them cannot be so understood. Therefore, by PP, Aristotle can claim that regardless of what else might be wrong with the arguments, the ultimate source of confusion is a failure to distinguish accidental from essential predications. Until that ontological distinction is pointed out and understood, complete resolution is impossible.

Aristotle, however, does not want to leave it that most or many of these examples have linguistic problems as well as the fundamental ontological problem. He also argues that examples such as the paternal dog *cannot* involve linguistic equivocation. He criticizes people who think that there are linguistic double meanings involved.

> Some people, too, resolve these reasonings by means of double meaning, for example, that he is your father, or son, or slave. And yet it is clear that if the refutation appears due to something being said in many ways, it is necessary that the name or phrase be said standardly (κυρίως) of more than one thing. No one standardly says that the child is his if he is a master of the child, but the combination is by accident. "Is this person yours?" "Yes." "And is this person a child? Then this person is your child since he happens to be both yours and a child." However he is not your child.[36]

Aristotle is somewhat careless here. He has forgotten that earlier in *S.E.* 4 he distinguished three ways in which double meaning can give rise to fallacy[37]:

1. when a name or phrase standardly (κυρίως) signifies more than one thing;
2. when a name or phrase customarily (even if not standardly) applies to more than one thing; and
3. when names singly signify one thing, but when combined result in a phrase with multiple signification.

Aristotle's point is that "child" (τέκνον) has only one standard signification, and he would have to add, it has only one customary signification. That is, the standard signification of "child" is the only signification. True, there are metaphorical uses of τέκνον: references, for example, to flowers as γαίας τέκνα (Aeschylus) or to birds as αἰθέρος τέκνα (Euripides). But at the core of these metaphors is the concept of generation or origin, not ownership. What Aristotle denies is any customary use of τέκνον to refer to one's property *qua* property. He denies, then, any possible case for double meaning playing a role in this fallacy.[38]

Whether or not we find this last argument convincing, there can be no doubt that the denial of purely linguistic resolutions of many of his examples is based upon Aristotle's Principle of Parsimony. Ignore or reject that principle, and there remains little reason why many of these fallacies must be resolved in the same way as the triangle argument.

HISTORICAL REASONS FOR TREATING FALLACIES DUE TO ACCIDENT AS ERRORS OF LOGICAL FORM

While the rejection of Aristotle's Principle of Parsimony allows treating many of these fallacies as generated by linguistic equivocation, it was the rejection of Aristotle's basic dichotomy between linguistic and nonlinguistic fallacies that led to the second non-Aristotelian interpretation of Accident, namely, that it is purely an error of logical form. To understand how this reinterpretation of Accident arose, a brief sketch of the medieval history of the fallacy is necessary.

Through the influence of Boethius, the fallacy of Accident became connected to a passage from *Categories* 3, which has some resemblance to the *S.E.* description of the fallacy. Aristotle has just distinguished two relationships in *Categories* 2: the "said of" relationship that applies species and higher genera to particulars under those universals (e.g., "animal is said of man"), and the "present in" relationship that applies nonsubstantial (i.e., accidental) predicates to substances (e.g., "white is present in Socrates"). In *Categories* 3, he remarks on the transitivity of the "said of" relationship:

> Whenever one thing is predicated of another as of a subject, all things said of what is predicated will be said of the subject also. For example, man is predicated of the individual man, and animal of man; so animal will be predicated of the individual man also, for the individual man is both a man and an animal.[39]

Boethius understood Aristotle to be saying that only in predications of the "said of" variety was it the case that whatever was true of the predicate also was true of the subject, and violations of that restriction produced errors due to Accident. In light of our analysis of Aristotle, what this Boethian rendering amounts to is a loosening of the requirement for double καθ' αὑτό belonging of the accidental premise. All that one required was a predication of a universal that belonged καθ' αὑτό to a particular under it. This interpretation, however, still supposed the error to be founded on the nature of things rather than on the multivocity of words. That remained the standard early medieval understanding of Aristotle's contrast between fallacies due to language (*in dictione*) and fallacies outside of language (*extra dictione*).

By the fourteenth century, however, there came to coexist with this Boethian interpretation a tradition of accounting for particular examples of Aristotle's fallacy by appealing to linguistic double meaning in one or more of the terms. This, we have seen, was a reasonable analysis of several of Aristotle's examples once PP was ignored. It was William of Ockham, in his *Expositio super libros elenchorum,* who attacked this tradition head on. One could not, he said, resolve Aristotle's examples of Accident linguistically, all the while claiming that fallacies *extra dictione* had their source in things rather than in words. To preserve the dichotomy between fallacies *in dictione* and fallacies *extra dictione*, William denied that fallacies due to Accident ever involved linguistic ambiguity. To account for the fallacies, however, he reinterpreted that basic dichotomy. The difference between *in dictione* fallacies and *extra dictione* fallacies was that the former arose in spoken or written speech, while the latter arose in an internal mental language. William's account of this mental language, which always preceded any instantiation in external speech or writing, was a subject of heated dispute in the late Middle Ages.[40] For our purposes we emphasize a couple of characteristics of this mental language. It was an ideal language, stripped of all the troublesome possibilities of ambiguity and equivocation found in spoken or written languages. There was an exact one-to-one correspondence between terms (signifiers) and things to be signified. It included as analogues to spoken languages only those grammatical or syntactical distinctions that were necessary for complete signification (*necessitas significationis*). That is, the criterion for inclusion as a component of mental language was "indispensability for description, [which] is applied by asking about any grammatical distinction whether its use makes a difference in the truth values of propositions."[41] Mental language was the same for all humans (as well as angels) and captured the formal structure of thought. It alone was the proper object of study by the logician, although insofar as spoken languages imperfectly instantiated the ideal mental language, they too might be investigated by the logician.

By transferring Aristotle's distinction between errors *in dictione* and errors *extra dictione* to this contrast between actual spoken languages and the ideal mental language, William of Ockham redefined fallacies due to Accident as errors of logical form. No longer was there any room for the Aristotelian or Boethian concern for the content of the predication (accidental or essential). Violations of valid logical form became the defining mark of Accident and other errors outside of language. Such a claim continues to resonate in modern treatments of these fallacies.[42] There was, too, another incentive for ignoring the ontological nature of the predications. Aristotle himself had described the error of one subclass of fallacies due to Accident in purely logical terms. He wrote that errors due to Consequent arose from a failure of "convertibility" in a premise. The natural way to interpret convertibility is as a formal operation relating logical terms or propositions. In the next section, I show that Aristotle's discussion of this subset of arguments does not support the interpretation of these failures as merely defects in logical form.

FALLACIES DUE TO CONSEQUENT

Introduction

In this section I analyze Aristotle's comments on and examples of fallacies due to Consequent (παρὰ τὸ ἑπόμενον). In particular, I try to make some sense of the distinction that Aristotle seems intent on drawing between other fallacies due to Accident and this particular subgroup. I conclude that no clear distinction emerges from his discussions and examples.

Aristotle introduces us to the fallacy in *S.E.* 5:

> The refutation due to Consequent is on account of thinking that the implication is convertible. For whenever it is true that since this is the case then that is necessarily the case, they think also that since that is the case then the first is necessarily the case.[43]

It arises because of a false belief that one can "convert" (ἀντιστρέφειν) the "implication" (ἀκολούθησις). Immediately we are confronted with two key terms with multiple Aristotelian uses. The verb ἀκολουθεῖν has at least four uses in Aristotle's logical and dialectical writings.[44] It might mean:

1. "to follow from," thereby signifying the relationship of logical consequence;
2. "to apply to," thereby signifying the relationship of one predicate holding of another;
3. "to be compatible with," thereby signifying the relationship of logical consistency; and
4. "to be the same as," thereby signifying the relationship of logical equivalence.

More important to note is that uses (1), (3), and (4) may designate relationships between propositions or between terms, while (2) only designates a relationship between the subject and predicate terms of a proposition. Analogously, the corresponding noun "implication" (ἀκολούθησις) can signify either a predicated *term* or an entire *proposition* (either implied by, compatible with, or equivalent to another term or proposition). This same ambiguity between terms and propositions arises in Aristotle's uses of "convertibility" (ἀντιστρέφειν).[45] Sometimes Aristotle will refer to the terms (ὅροι) of a proposition being convertible. Other times convertibility will apply to propositions (προτάσεις).[46] The problem for the reader of *S.E.* is how to construe "convertibility of an implication" as an intended analysis of the fallacy due to Consequent.

Let us look again at how Aristotle describes the error made by those who reason falsely due to Consequent.

> For whenever it is true that since this (τόδε) is the case then that (τόδε) is necessarily the case, they think also that since that (τόδε) is the case then the first is necessarily the case.[47]

If the error applies to terms, the false belief is that "Every A is C" is convertible with (i.e., implies, is consistent with, or is equal to) "Every C is A." On the other hand, if the error applies to propositions, the false belief is that "If A then C" is convertible with (i.e., implies, is consistent with, or is equal to) "If C then A." Schematically, this propositional form of the error might seem initially to be:

	If A then C.
Therefore,	if C then A.

But if the indefinite placeholders (τόδε) refer to propositions, the schema requires expansion. By this interpretation of convertibility, if premise A entails conclusion C, then conclusion C also entails premise A. In both cases the move from one premise to a conclusion requires a second premise. So the schema for premise-to-conclusion convertibility actually looks like this:

If	A	Then	C
	B		B
Therefore,	C	Therefore,	A

What makes the second syllogism a case of misreasoning is that B is an accidental predication. What makes that same syllogism persuasive is that the same accidental predication did not invalidate the first syllogism. Its harmlessness in the first argument, the sophist might claim, carries over to the second.

Are we to understand the false belief in convertibility that characterizes Aristotle's examples of Consequent to apply to terms or to propositions?

Despite hopes that Aristotle's examples would decide the issue, I find that most of his examples are too abbreviated not to admit of more than one interpretation, and that neither interpretation can account happily for all of the examples.

In what follows I analyze Aristotle's examples of false arguments due to Consequent in both ways. That is, I show how the error might trade (1) upon the false belief in convertibility between propositions, specifically, between a premise and the conclusion of a valid argument, or (2) upon the false belief in term convertibility of a universal accidental premise. I argue that a false belief in premise-to-conclusion convertibility is the best interpretation to provide a real difference between Consequent and other forms of Accident. It explains how one might be persuaded to accept a false syllogism apart from simply misidentifying an accidental predication as an essential predication. Term convertibility, on the other hand, seems hardly distinguishable from other errors of Accident. But as useful as premise-to-conclusion convertibility is for many of Aristotle's examples of Consequent, it fails to account for his favorite examples of such misreasoning where simple term convertibility seems to be the error in question. I then analyze Aristotle's own problematic explanation of how Accident and Consequent differ and conclude that the distinction is ultimately unjustified and philosophically useless.

Aristotle's Examples

Aristotle begins with a couple of common, everyday examples illustrating the sort of deception that arises by Consequent. Because honey is yellow, he says, people are fooled into believing that gall is honey because it is yellow. Aristotle says that this particular deception involving honey and gall occurs "often" (πολλάκις).[48] It is hard to imagine such a mistake commonly arising among mature Greeks. He may be thinking of the standard practice among Greek mothers and wet nurses of putting honey on their nipples to encourage suckling, and then changing from honey to gall when the child was to be weaned off the breast.[49] Aristotle seems to have thought that the honey-like color of the gall deceives the child to be weaned. Perhaps, too, this is why Aristotle does not say that this particular deception arises from false reasoning (συλλογισμός), but that it is "an opinion arising from sense perception."[50] Infants would be vulnerable to the latter, not the former. But even if the infant is not engaged in false reasoning *per se*, the deception admits of syllogistic analysis. If ἀκολούθησις refers to a relationship between propositions, then Aristotle may have in mind two syllogisms, one valid and one invalid.

[A] (1) This thing is honey.	[B] (3) This thing (i.e., gall) is yellow.
(2) All honey is yellow.	(2) All honey is yellow.
(3) This thing is yellow.	(1) This thing (i.e., gall) is honey.

The error is to think that because in [A] the conclusion (3) necessarily follows from premise (1) (by the predication (2)), that therefore in [B] premise (1) necessarily follows from the conclusion (3) (again by the same predication (2)). The question, then, is why [B] fails. The answer is that it fails in the same way that all fallacies due to Accident fail. In the second premise, being honey and being yellow are not "indistinguishable and one in substance." If they were, then the argument would be sound.

Alternatively, if ἀκολούθησις refers to a relationship between terms of a predication, the error in the second syllogism is to believe that "All honey is yellow" (premise (2)) is convertible with "All yellow things are honey." The real problem with [B] as it stands alone, then, is that it is a classic case of mistaking premise (2) to be a nonaccidental predication. Nonconvertibility of terms is simply the formal symptom of the underlying mistake of thinking an accidental predication to be an essential predication. Why then should Aristotle wish to distinguish [B] from any other case of Accident? The traditional answer to this is that the accidental premise in [B] is universal, whereas accidental premises that characterize other fallacies due to Accident are particular. I present the difficulties with such a distinction below.

Aristotle's second example is similarly abbreviated. Because whenever it rains the earth becomes wet, people erroneously argue that if the earth has become wet, then it must have rained.[51] In its expanded form, as an error in premise-to-conclusion convertibility, the reasoning might begin with this valid argument:

(1) This ground has received rain.
(2) When it rains the ground becomes wet.

Therefore, (3) This ground is wet.

People then assume that the implication from premise (1) to conclusion (3) is convertible, and reason:

(3) This ground is wet.
(2) When it rains the ground becomes wet.

Therefore, (1) This ground has received rain.

Once again, appeal to Consequent accounts for the form of the false argument, but it does not explain why that form is fallacious. One must resolve the error of the second syllogism by explaining the accidental nature of the second premise. Alternatively, the error might just consist in believing that the universal premise (2) is convertible. We have then the same problem: the reason it is not convertible is because it is accidental, and the term convertibility analysis seems to render this indistinguishable from any other fallacy due to Accident.

Aristotle likens these mistakes to false rhetorical arguments based upon signs. In both the *Rhetoric* and the *Prior Analytics,* valid and invalid arguments from signs are distinguished according to the necessity of the relationship between sign and signified.[52] An orator might argue, for instance, that someone is an adulterer because he adorns himself and wanders about at night, since that is what adulterers do.[53] But even granting that all adulterers do such things, it is only a probable sign rather than a necessary sign, since not all people who do such things are adulterers. Traditionally, these arguments are thought to founder on an illegitimate term conversion of the second premise: all adulterers are nighttime wanderers, therefore, all nighttime wanderers are adulterers. The orator (or sophist), however, would not wish to make explicit the claim that all nighttime wanderers are adulterers, for he would expect such a claim to be rejected by his audience. Instead, he might hide the universal in the interest of arguing that some particular nighttime wanderer is an adulterer. In this way, such a rhetorical argument could possess the same form as Aristotle's earlier examples of false premise-to-conclusion convertibility.

(1) This person is an adulterer.
(2) All adulterers are nighttime wanderers.
____
(3) This person is a nighttime wanderer.

In the belief that the implication is convertible, one might then reason:

(3) Socrates is a nighttime wanderer.
(2) All adulterers are nighttime wanderers.
____
(1) Socrates is an adulterer.[54]

Once again, although the fallacy is due to Accident (for nighttime wandering does not belong καθ' αὑτό to being an adulterer), the persuasiveness of the false syllogism arises because of the validity of the first syllogism.

I have tried to make a case that what distinguishes a fallacy due to Consequent from other forms of fallacious reasoning due to Accident is that one is led to commit the fallacious reasoning because of a belief in the convertibility of the implication from some other valid syllogism. The failure of "conversion of implication" (i.e., Consequent) explains how one might be led into a piece of false reasoning, but it does not explain why that reasoning is false. Accident accounts for the actual falsehood. This interpretation has the advantage of offering a real difference between these examples and Aristotle's simple fallacies due to Accident. But premise-to-conclusion convertibility fails to account for the most prominent illustration of the fallacy: the argument of Melissus for the spatial infinity of the universe. Among serious philosophical errors, this argument ranks among Aristotle's favorite whipping boys. It is attacked three different times in *S.E.* as well as in the *Physics*.[55] The argument is first mentioned as an example of a "syllogistic" (συλλογιστικός) fallacy due to Consequent in contrast to the rhetorical (ῥητορικός) examples cited earlier.

> In the argument of Melissus that the universe is infinite, the fallacy arises by accepting on the one hand that the universe is ungenerated (for from what is not, nothing can come to be) and on the other hand that what has been generated was generated from a beginning. So if it has not been generated [it is necessary that] the universe does not have a beginning, so that it is infinite. But it is not necessary that this be so. For it is not the case that if all that is generated has a beginning, then also if something has a beginning then it has been generated, just as it is not the case that if one with fever is hot, then too, it is necessary that the one who is hot is with fever.[56]

The argument of Melissus seems to be:

(1) The universe is ungenerated.
(2) What has been generated has a beginning.

---

(3) The universe does not have a beginning.

Aristotle notes that the truth of (2) does not imply its converse. The fact that all generated things have beginnings does not entail that all things with beginnings have been generated, any more than the fact that all persons with fever are hot entails that all hot persons are fevered. But this appears to be a simple case of erroneously believing in the term convertibility of the premise. It is not, however, direct term convertibility (i.e., between "Every A is B" and "Every B is A"). Instead, it is what Aristotle calls "convertibility by opposites" (i.e., between "Every A is B" and "Every non-A is non-B"). This is made clear in *S.E.* 28:

> The implication from Consequent is possible in two ways. For either as the universal is implied in a part, e. g., as animal is implied in man (for they expect that if this occurs along with that, then that also occurs along with this), or else the implication arises by opposites (for if this is implied in that, then the opposite [of this] is implied in the opposite [of that]. And the argument of Melissus is due to this, for if that which has come to be has a beginning, he expects that that which is ungenerated does not have a beginning, so that if the heaven is ungenerated it is also infinite. But it is not so, for the implication is in the opposite direction.[57]

The question remains, if convertibility refers to the terms of a premise, and the error arises from the false belief that accidental predications are convertible (either directly or by opposites), why distinguish Consequent from Accident? Aristotle's only justification of the distinction is very difficult to interpret. I quote it in its entirety.

> Fallacies due to Consequent are a part of those due to Accident. For the consequent just happened [to follow]. But it differs from Accident since it is possible to obtain the fallacy of Accident in the case of only one (ἐφ' ἑνὸς μόνου), e.g. [to conclude that] the yellow thing and honey are the same, and the white thing and a swan, but the fallacy due to Consequent always arises in more than one thing (ἀεὶ ἐν πλείοσι). For things that are the same as one and the same thing, we also deem to be the same as each other. Wherefore a refutation arises due to Consequent. But it is possible for it to be not altogether true, e.g., if they were the same by accident. For both the snow and the swan are the same as something white. Or again, as in the argument of Melissus, he takes it that "to have been generated" and "to have a beginning" are the same. . . . For since what has been generated has a beginning, he also deems that what has a beginning has been generated, as both what has been generated and what is finite are the same as what has a beginning.[58]

What are we to make of the distinction between Accident that arises ἐφ' ἑνὸς μόνου and Consequent that arises ἀεὶ ἐν πλείσι? The traditional understanding of this dichotomy is that it distinguishes between a particular and a universal premise.[59] By this interpretation, fallacies due to Accident involve mistaking a *particular* accidental predication for a *particular* essential predication; fallacies due to Consequent involve mistaking a *universal* accidental predication for a *universal* essential predication. This reading has two advantages:

1. It connects the distinction to term convertibility. Recall that convertibility was never mentioned in Aristotle's discussions of Accident but became the prominent notion in his treatment of Consequent. Whereas universal accidental premises are never truly convertible, particular premises sometimes are (e.g., "Coriscus is the one approaching").[60]
2. Almost all of Aristotle's examples of Accident involve particular accidental premises, and all of his examples of Consequent involve universal accidental premises.

There are, however, difficulties with this interpretation. First, at least one example of Accident has a universal premise. The *S.E.* 6 triangle fallacy hinges on the premise: "All triangles are figures."[61] Moreover, this triangle argument is cited as illustrating the fallacy of Accident within twenty lines of Aristotle making the problematic distinction between Accident and Consequent. It seems improbable that Aristotle would specify a distinction immediately after violating it.

Second and most damaging to the traditional interpretation is that neither the immediate explanation of the distinction nor the examples chosen to illustrate the distinction seem to have anything to do with the particular/universal dichotomy. In the text above, the justification offered for saying that Consequent arises ἀεὶ ἐν πλείοσι is that:

> things that are the same as one and the same thing, we also deem to be the same as each other. Wherefore a refutation arises due to Consequent.

In form the error seems to be:

(1) A is [the same as] C.
(2) B is [the same as] C.

Therefore, (3) A is [the same as] B.

where (3) is false because (1) and (2) are accidental predications.[62] The two examples that follow confirm this suggestion:

	(1) Snow is white.		(1) What is generated has a beginning.
	(2) A swan is white.		(2) What is finite has a beginning.
Therefore,[63]	(3a) A swan is snow.	Therefore,	(3a) What is generated is finite.
	(3b) Snow is a swan.		(3b) What is finite is generated.

Contrast the above schema for Consequent with the examples cited for Accident, which arise ἐφ' ἑνὸς μόνου:

(A) "the yellow thing is [the same as] honey."
(B) "the white thing is [the same as] a swan."

Aristotle here singles out the premise that is mistakenly thought to be nonaccidental in the course of two fallacious arguments. But the example of yellow and honey, here used as an example of Accident to *contrast* to Consequent, was cited in *S.E.* 5 as an example of Consequent, not Accident!

What are we to make of all this? Is the difference between Accident and Consequent the difference between misinterpreting a universal or a particular accidental predication? *S.E.* 6 offers no suggestion of this and violates it explicitly with the triangle argument. My own conclusion is that Aristotle either (1) had no clear distinction in mind between Accident and Consequent (witness his use of "the yellow thing is honey" predication first as an example of Consequent in *S.E.* 5, then as an example of Accident in *S.E.* 6), or (2) had some distinction in mind that he never properly exemplified. What that distinction might have been is hard to say. Maybe it was that between a universal and a particular accidental premise. On the other hand, maybe we are expected to fill out the phrases ἐφ' ἑνὸς μόνου and ἀεὶ ἐν πλείοσι with συλλογισμός as the missing reference, thus to read ἐφ' ἑνὸς μόνου συλλογισμοῦ and ἀεὶ ἐν πλείοσι συλλογισμοῖς. If so, then Accident arises by one syllogism alone, whereas Consequent always arises through more than one syllogism. This would fit with the interpretation of Consequent as applying to premise-to-conclusion convertibility, where a prior valid syllogism smoothes the way toward accepting the subsequent false syllogism. It also might account for the *S.E.* 6 analysis of Melissus's argument, where one syllogism is used to (falsely) derive "what is generated is finite," or "what is finite is generated," and a second syllogism uses that predication to derive (validly or invalidly) Melissus's false conclusion about the infinity of the universe. This, however, can only be speculation. The text as we have it, with the examples that Aristotle provides, offers us a distinction of little use, either to Aristotle or to us.

Conclusion

Despite Aristotle's attempts to mark off fallacies due to Consequent from fallacies due to Accident, he seems to leave us with a single type of false reasoning calling for a single type of resolution. Although these arguments involve three terms and admit a formal analysis, Aristotle never appeals to logical form to explain the error. Alternatively, one might account for these false appearances of formal validity by citing linguistic equivocation. Yet Aristotle explicitly denies that verbal ambiguity is the source of the error. Rather, in each example he blames the accidental nature of one of the premises as the cause of the misreasoning. The victim falsely supposes that the terms of the premise signify things "indistinguishable and one in substance." Such a predication amounts to that double καθ' αὑτό belonging, specified in *Pst. An.* I, 4: the terms must signify things that are essential to one another.

Aristotle considers a number of different resolutions to his examples of false reasoning due to Accident. Each alternative is rejected by appeal to the Principle of Parsimony. Whereas particular examples will admit different resolutions, only the ontological error of mistaking an accidental predication for a predication of identity can account for all of the examples. Therefore, the only proper resolutions *to the arguments* require reference to Accident, even if other resolutions may be effective against different arguers. These errors are outside of language, because no amount of linguistic clarification is sufficient protection against them. One also needs proficiency in knowing when things are "one in substance." This requires of the successful dialectician a grasp of the essences of things. He must know when the subjects and predicates of his premises belong to the essences of each other.

Chapter 8

Secundum Quid

Introduction

The next fallacy outside of language is identified in *S.E.* by several cumbersome phrases: "due to something being said either without qualification, or in some way and not standardly" (*S.E.* 5), "due to something being said either standardly, or in some way or some place or some respect or relative to something and not without qualification" (*S.E.* 25). The most abbreviated form is simply "due to [something being said] in some way and without qualification" (παρὰ τὸ πῇ καὶ ἁπλῶς), which was rendered by the Latin "secundum quid et simpliciter." I will use the abbreviated Latin *Secundum Quid* as a convenient label for this family of errors.

> Fallacies due to something being said either without qualification (ἁπλῶς) or in some way (πῇ) and not standardly (κυρίως) arise when what is said in part (ἐν μέρει) has been taken as having been said without qualification.[1]

Aristotle here contrasts what seem to be two ways of speaking but are actually two ways of being, signified by slight differences in the signifying phrases. On one side of the contrast stand the adverbial words "without qualification" (ἁπλῶς) and "standardly" (κυρίως); on the other side stand "in some way" (πῇ) and "in part" (ἐν μέρει). Of all Aristotle's fallacies outside of language, this error is most easily construed as a purely linguistic confusion. One modern scholar has argued at some length that Aristotle has misclassified this error on the wrong side of his dichotomy between linguistic and nonlinguistic errors.[2] Although Aristotle is aware of the temptation to analyze these fallacies as linguistic equivocations, he takes pains to deny that such an analysis is correct. But he admits that language is partially to blame for the deceptiveness of such false arguments, and some linguistic clarification is necessary for a complete resolution. And if the criterion that sets apart fallacies due to language from fallacies outside of language is the mere presence of some false presupposition about language or its relationship to reality that makes the argument appear sound, then *Secundum Quid* should be considered as much a fallacy due to language as any of those so classified.

Aristotle's analysis of *Secundum Quid* fallacies is similar to his analysis of Form of the Expression fallacies. Both derive their persuasiveness from false presuppositions both about language and about what language signifies. If the *necessity* of some linguistic clarification for resolution renders a fallacy type "due to language," then *Secundum Quid* and Form of the Expression are both due to language. If the *sufficiency* of some linguistic clarification for resolution renders a fallacy type "due to language," then neither *Secundum Quid* nor Form of the Expression are due to language. What Aristotle fails to justify is his classifying one type (Form of the Expression) due to language and the other (*Secundum Quid*) outside of language.

TWO TYPES OF *SECUNDUM QUID* FALLACY

We can divide *Secundum Quid* fallacies into two types: those that confuse existential with predicative modes of being and those that confuse two different modes of predicative being. What Aristotle insists upon distinguishing, if one is not to be deceived by these arguments, are three states of being:

(a) x existing without qualification,
(b) x being f without qualification, and
(c) x being f in some qualified way.

The deceptions arise because such different states of being are not always clearly signified in speech.

The first type of *Secundum Quid* fallacy involves a confusion between (a) and (b) above. We might call this a confusion of the existential "is" and the predicative "is." These are arguments claiming either that x is (i.e., exists) because x is f or, conversely, that x is not (i.e., does not exist), because x is not f, for some predicate f. Any existing thing can be the subject of the latter version of this type, since of any existing thing there are innumerable predications that cannot truly be made (e.g., trees are not persons, therefore trees are not, i.e., do not exist). The former version arises in special cases where words or phrases signify to us but do not signify by nature.[3] For example,

> if what-is-not is an object of opinion, one concludes that what-is-not is.[4]

Contrary to the Parmenidean tradition, Aristotle acknowledges that one can make significant predications of what does not exist, as long as what does not exist has signification to us.[5]

Aristotle describes the ontological error underlying these fallacies:

> For to-be-something and to-be without qualification are not the same . . . [and] not-to-be-something and not-to-be without qualification are not the same.[6]

Equally succinct is the account of why someone might make the above mistakes:

> They appear to be the same because of the close resemblance of the language and because "to-be-something" is little different from "to-be," and "not-to-be-something" is little different from "not-to-be."[7]

To appreciate the full resolution of this error, one must acknowledge the connection between these two passages without blurring the distinction. The false presupposition itself is ontological. The believability of the presupposition (i.e., its appearance of truth) is promoted by the additional false belief that small differences between linguistic descriptions reflect negligible differences in the states described. The particularly close connection between language use and ontological confusion in this fallacy accounts for much of the criticism that Aristotle has received for classifying this fallacy as one outside of language. I discuss this relationship below and in Appendix 4.

The second type of the *Secundum Quid* fallacy involves not a confusion between existential being and predicative being but between two kinds of predicative being: either between qualified and unqualified f-ness, or between two differently qualified ways of f-ness. Aristotle's classic example of a confusion between being f in a qualified and in an unqualified way is the Indian (or sometimes, the Ethiopian) who is white with respect to his teeth but dark, that is, not white, unqualifiedly. From this it is sophistically concluded that he is both white and not white. Aristotle doubts that many of these arguments truly deceive. More dangerous are cases of two qualified predications, neither of which admit of unqualified predication. For instance, if something is half white and half black and there is no more reason to predicate one instead of the other without qualification, one feels compelled either to claim that both contraries apply without qualification or that neither do. The first option results in real contradiction: if both apply, then the object is both unqualifiedly f and unqualifiedly not-f. If neither applies, then the object is not unqualifiedly f and not unqualifiedly not-f. This, however, is only an apparent contradiction. For the denial of not being unqualifiedly f is being unqualifiedly f, which is not the same as not being unqualifiedly not-f.

RESOLUTIONS OF *SECUNDUM QUID* FALLACIES

Sorting out real from only apparent denials is the first step in resolving these false reasonings. Failure to do so is evidence of ignorance of what a refutation is.

> Fallacies due to predicating in some way and without qualification [are reducible to *Ignoratio Elenchi*] since the affirmation and the denial are not of the same thing. For the denial of being-white-in-some-way is not-being-white-in-some-way, and the denial of being-white-without-qualification is not-being-white-without-qualification. If then, upon granting that something is white in some way, he takes it as being said without qualification, he does not produce a refutation, but appears to on account of ignorance of what a refutation is.[8]

This recalls Aristotle's stipulated requirements for real refutation. There must be

1. premises that do not include the conclusion;
2. a conclusion that follows necessarily from the premises; and
3. a conclusion that denies one and the same predicate affirmed by the answerer. That predicate denied must be: the thing signified, not just the name; (3b) the thing under the same description as affirmed by the answerer; and (3c) the thing qualified in precisely the same way as it was affirmed by the answerer.

False refutations due to *Secundum Quid* violate (3c):

> For contraries, and opposites, and an affirmation and a denial, cannot without qualification belong to the same thing. But nothing prevents both belonging in some way or in some relation or in some respect, or one belonging in some way while the other belongs without qualification. As a result, if one thing belongs without qualification and one thing belongs in some way, there is not yet a refutation, and this is why one must examine the conclusion relative to its denial.[9]

Secundum Quid As a Fallacy outside of Language: Aristotle's Position

The question that Aristotle's account of *Secundum Quid* fallacies raises is why should these not be classified as due to language? That Aristotle rejects such an analysis is not disputable, but is that rejection consistent with the rest of Aristotle's analysis of fallacies? Kirwan claims that one of the ways Aristotle's classification of fallacies goes astray is by not recognizing that errors due to *Secundum Quid* are errors due to double meaning, such as homonymy and amphiboly.[10] According to Kirwan, Aristotle's examples of *Secundum Quid* are examples of what he calls "equivocal elision." For example, in the classic Indian or Ethiopian argument, "white" equivocates between "white-skinned" and "white-toothed." Basically Kirwan considers *Secundum Quid* fallacies as instances of amphiboly, where the syntax of a sentence induces a double meaning of some otherwise univocal word or phrase.

Aristotle is aware that these fallacies sometimes look like mere linguistic confusions. We know this from two comments he makes, one in his treatment of *Ignoratio Elenchi* and the other in the chapter on Accident. In *S.E.*, 5 Aristotle first introduced the reader to the possibility of accounting for all of the kinds of false reasonings under the heading "ignorance of what refutation is" (*Ignoratio Elenchi*). It was here that the definition of a refutation was expanded to show how each type of error violated one or more of the clauses of the definition. The only examples offered in that passage of the failure to abide by all of these requirements involved the neglect of the requirement that the conclusion deny in precisely the same way (i.e., with the same qualifications) what the answerer earlier affirmed. These were sophistic arguments to the conclusion that the same thing is both double and not double.[11] For instance, one might fallaciously conclude that something can be both double and not double by reasoning that two things are both double of one thing and not double of three things. This, however, does not refute the claim that nothing can be both double and not double, because the two things are not both double and not double "of the same thing" (κατὰ ταὐτό). Or, again, one might reason that A is both double of B in length and not double of B in breadth. Again, it is a false refutation, says Aristotle, for although A is both double and not double κατὰ ταὐτό, it is not both double and not double in the same respect or in the same way. Or one might show one thing being both double and not double of one and the same object and in the same way, yet not at the same time. Again, the result is only an apparent refutation. Aristotle then closes his discussion of *Ignoratio Elenchi* with the important but problematic statement:

> Someone might forcefully drag (ἕλκοι) this [fallacy] also into those due to language.[12]

This statement raises two questions. First, is Aristotle affirming or denying that this fallacy could properly be understood as due to language? Second, what exactly is the fallacy to which he is referring? That is, what is the antecedent of "this" (τοῦτον) in the claim? To answer the first question, Poste seems correct in understanding Aristotle not to be conceding a possible alternative analysis of the fallacy. The use of ἕλκω suggests some degree of violence being done by the alternative analysis.[13] Aristotle, then, is here reporting critically, and not endorsing, the reduction of this fallacy to one due to language. But what is "this fallacy"? There are two defensible antecedents for τοῦτον.

1. It usually is taken to refer to the entire class of fallacies due to *Ignoratio Elenchi*.[14] On this reading, the analysis being rejected by Aristotle amounts to reducing all fallacies to linguistic sources. Aristotle will go on to show in *S.E.* 6 that all of the sources of false reasoning, both linguistic and nonlinguistic, can be understood as due to *Ignoratio Elenchi*. The rejected analysis, then, is one that denies the linguistic versus nonlinguistic dichotomy underlying Aristotle's entire treatment.
2. The fallacy that someone might "drag" under the head "due to language" may only be the particular fallacy illustrated by the "double and not double" examples. In other words, the reference may be just to fallacies of *Secundum Quid*. On this reading, the rejected analysis does not involve a general reduction of all fallacy to linguistic confusion but the reduction of one sort of (nonlinguistic) fallacy to linguistic confusion.[15]

Whether Aristotle intends the rejected analysis here to be the universal reduction of all false reasoning to linguistic confusion, or a more local reduction of *Secundum Quid* fallacies to linguistic confusion, is difficult to determine, since both analyses are elsewhere rejected. My own belief is that it is more natural to read Aristotle as referring to the fallacy that he has just exemplified, that is, to *Secundum Quid*. If so, it reveals his particular sensitivity to the appearance of double meaning in just these sorts of errors.

The same sensitivity is evident in a second text at the end of the *S.E.* 24 treatment of proper and improper resolutions of false refutations due to Accident. There Aristotle directly addresses the precise charge of Kirwan, that expressions that may be understood either as qualified or unqualified are cases of double meaning by ellipsis. After explaining why certain fallacies due to Accident cannot be resolved by appealing to double meaning, Aristotle considers a new piece of sophistical reasoning that hinges on what seems to be an equivocation between rendering the genitive case as partitive or possessive. His point is that these are not cases of linguistic ambiguity but cases of *Secundum Quid.* The argument is difficult to render into uninflected English. The sophistical conclusion is that "some evil is good" (τὸ εἶναι τῶν κακῶν τι ἀγαθόν). It is reached by the premises:

1. Prudence is good.
2. Prudence is the knowledge of evil. (ἡ φρόνησις ἐστιν ἐπιστήμη τῶν κακῶν.)

Therefore something of evil (i.e., knowledge) is good,

which is another possible way of reading τὸ εἶναι τῶν κακῶν τι ἀγαθόν. The resolution rejected by Aristotle is that the genitive case ("this is of that": τοῦτο τούτων εἶναι) is "said in many ways." On the contrary, he says, the genitive case signifies one thing: possession.[16]

Aristotle knows that this hardly settles the issue, and he begins a dialectical exchange with himself to further clarify the issue. Let us even grant to someone, he suggests, *contrary to what we really believe*,[17] that the genitive is an example of double meaning, because of this argument:

> For we say that "man is of the animals" [= "man is an animal," partitive genitive] but not that he is some possession of the animals; and even if something that is related to evils is described as "of something," it is because that thing is related to evils, not that it is one of the evils.[18]

Even if one were to concede this argument for the double meaning of the genitive case, Aristotle's final conclusion is that "it has this appearance due to *Secundum Quid.*"[19] So the appearance of double meaning is due to one's confusing the unqualified sense of the genitive (possessive) with the genitive qualified in some way (partitive or relative). But Aristotle continues to ponder the appearance of equivocation:

> Nevertheless, perhaps it is possible to say "good is of evils" in two ways, but not in the case of this argument. A more likely [case would be] if some slave were good of the wicked [= some good slave belonged to a wicked man]. But perhaps there is no double meaning here either, for it is not the case that if a slave were good and his [lit. "of him"] he would at the same time be "his good."[20]

The best cases to be made for double meaning of ἀγαθὸν εἶναί τι τῶν κακῶν are arguments that make use of the fact of a good slave being the possession of a bad man. Then one could truly say that there was a certain good (slave) of a bad (master), which also might be misunderstood to mean that a certain good was one of the bad (partitive genitive). The case for double meaning, then, is that a phrase such as ἀγαθὸν εἶναί τι τῶν κακῶν is amphibolous between A *belongs to* B, and A *is one of the* B's. But Aristotle is not willing to grant even this as a proper resolution.[21] He cites the real error as due to Accident. It is akin to the "paternal dog" argument[22]:

(1) That dog is yours.	(1) That slave is his.
(2) That dog is a father.	(2) That slave is good.
(3) That father is yours.	(3) That good is his.

Just as the reasoning from "He is a good slave of a bad man" to "The bad man is good" confuses an accidental predication for an essential identity and is not an example of linguistic equivocation,

> nor is saying "man is of the animals" said in many ways. For it is not the case when we signify something by separating off [an accident] that it is said in many ways. For we even signify "Give me the *Iliad*" by saying the half-line, i.e., "Give me 'Sing, goddess, the wrath."[23]

When we signify the *Iliad* by citing a half line, we are signifying by separating off (ἀφελόντες) a particular part or feature of the whole. So in predicating of the *Iliad* that it is "Sing, goddess, the wrath," we are not assuming that the *Iliad* is said in many ways. Likewise, when we predicate of man that he is "of the animals," we are signifying man by separating off an accident of man. (Accident, recall, is being contrasted to identity, i.e., with double καθ' αὑτό belonging.) Aristotle's point seems to be that man is not said in many ways any more than the *Iliad* is, just because it can be signified ἀφελόντες, that is, signified by reference to some nonidentical characteristic. The unabbreviated formulation for man would be an Aristotelian definition (e.g., "rational biped animal"). Aristotle's answer to Kirwan, then, might be that if "equivocal elision" were allowed as a form of linguistic ambiguity, then every subject that admits accidental properties is ambiguous, and that amounts to the claim that there are no univocal subjects.[24]

PROBLEMS WITH ARISTOTLE'S POSITION

We have seen that Aristotle recognizes the appearance of equivocation in fallacies due to *Secundum Quid.* We also noted that with fallacies due to

Accident the appearance of verbal equivocation was strong in some examples, but nonexistent in, for example, the triangle argument. By appealing to the Principle of Parsimony, the fact that some fallacies due to Accident did not involve any possible linguistic multivocity was sufficient for Aristotle to exclude all of the examples from requiring linguistic clarifications in their resolution. But PP is unable to perform the same job in resolving errors due to *Secundum Quid*. These false appearances are not merely artifacts of several examples of the fallacy. The appearance of verbal equivocation in these examples seems to be intrinsically bound to the nature of the fallacy itself, not just to selected examples of the fallacy. One reason for this can be traced, I believe, to an attempt by Aristotle in the *Topics* and *S.E.* to reinterpret what was once merely a linguistic distinction as one having full ontological signification. I argue in Appendix 4 that ἁπλῶς in pre-Aristotelian Greek was reserved exclusively for descriptions of a way of speaking. It meant saying something simply, without any additional words. I claim that Aristotle supplants that use of ἁπλῶς by κυρίως and attempts to use ἁπλῶς in a new fashion to distinguish between ontological conditions. But even in Aristotle the linguistic heritage of ἁπλῶς is never so entirely erased that he can deny some causal role of ordinary speech in the production of *Secundum Quid* fallacies. Although ways of speaking cannot affect ways of being, they do affect our beliefs about the way things are. Therein lies the danger of *Secundum Quid* fallacies. Just as Aristotle blamed the miniscule linguistic difference between τὸ εἶναί τι and τὸ εἶναι as sometimes responsible for the confusion of predicative and existential being, the same seemingly insignificant linguistic differences can lead one into other *Secundum Quid* confusions.

> And with fallacies that are due to predicating in some way and without qualification, the deception resides in the smallness [of the difference]. For we accept the predication universally, as though predicating something in some way or in some respect or at this time has no additional signification.[25]

What we have then is ontological confusion engendered by deceptive ways of speaking. There exists in the victim a false presupposition about how words relate to things,[26] which must be corrected for complete resolution. One removes the perplexity that arises from such sophisms by showing that minuteness of verbal differences in predications does not reflect insignificant differences in the things signified. As a result, the exact predication intended by the answerer is not denied in the same way by the questioner. As for the further mystery of why one thought it was being denied, that is resolved by distinguishing different ways of being. In these respects then Aristotle's account of fallacies due to *Secundum Quid* is not dissimilar to his account of fallacies due to Form of the Expression. And if Form of the Expression is to be regarded as a linguistic fallacy because some false presupposition about language requires correction, the same classification should be made for *Secundum Quid.*

There are two other reasons why fallacies due to *Secundum Quid* are closely associated with fallacies due to Form of the Expression and should be classified with them as fallacies due to language. First, we noted in chapter 3, pp. 40ff., that among Aristotle's examples of false reasoning due to Form of the Expression were confusions of substances with relatives. There Aristotle regarded qualifications of manner (ὡς, ὧδε) as signifying relatives (πρός τι), thereby qualifying as Category mistakes. In the expanded description of *Secundum Quid,* Aristotle contrasts ἁπλῶς predication with predication "in some way, or some place, or some respect, or relative to something."[27] At least two of these (place and relative) fall among the Categories.[28] It is difficult to see how such errors differ in kind from earlier examples of Form of the Expression, particularly if the manner of resolution determines the fallacy type. Both *Secundum Quid* and Form of the Expression require correction of both linguistic and ontological mistakes for complete resolution.

Second, Aristotle says in *S.E.* 6 that

> all the types [of fallacy] fall under *Ignoratio Elenchi*, those due to language because the denial (which is characteristic of a refutation) is only apparent, and the others on account of the definition of reasoning.[29]

The definition of reasoning he has in mind is that offered in the *Topics* and repeated in *S.E.*: an argument in which, when certain premises are set down, something different from the things set down follows by necessity through the things set down.[30] Excepting examples of *Secundum Quid*, all of the fallacies outside of language violate this definition. Begging the Question does not lead to something *different from* the premises set down. Non-Cause As Cause does not lead to something *through* the things set down. Accident and Consequent do not lead to something *by necessity*. Many Questions (I will argue in the next chapter) fails to set down *certain premises* in their proper form. On the other hand, all of the fallacies due to language violate in some way the stipulation that a proper refutation must lead to "a conclusion that denies one and the same predicate affirmed by the answerer." This also is what *Secundum Quid* fallacies violate.

CONCLUSION

We have found in Aristotle three kinds of false presupposition, the correction of which is necessary for the resolutions of false refutations. The first kind is a false belief about language, either an error as to what constitutes a linguistic signifier or an error as to the univocity of signifiers. The second is a false belief about the extralinguistic world itself. These include mistakes about the relative explanatory powers of things and events and mistakes about the different kinds of predication: differences in Categories of predicates, differences between essential and accidental predicates, and differences between qualified and unqualified ways of being.

There is a third kind of false presupposition: a mistaken belief about the way names and expressions relate to the things they signify. Such a misunderstood relationship amounts really to a combination of some error about language and some error about ontology. Fallacies due to Form of the Expression and fallacies due to *Secundum Quid* both derive their appearance of soundness from such a mistake about the relationship of language to reality. Because Aristotle classifies fallacy types according to the manner of their resolution, these two fallacy types should fall on the same side of the divide between linguistic and extralinguistic. The consistency of his taxonomy, however, breaks down. Perhaps when he analyzed *Secundum Quid* fallacies, Aristotle was more impressed by the ontological error of confusing qualified with unqualified states of being than by the linguistic error of believing that minor differences in description correlated to minor (i.e., negligible) differences in the events described. And perhaps when looking at Form of the Expression fallacies, the false belief that morphological or syntactical similarities among words correlated to ontological similarities in the things signified may have struck him more forcefully than the failure to recognize different Categories of predicates. In both cases, however, correction of the linguistic error presupposes prior understanding of the ontological distinctions in question. If Aristotle's position is that fallacy types due to language are characterized by some false presupposition about language that is necessary for the appearance of proper reasoning, then false arguments due to *Secundum Quid* must be classified as fallacies due to language.

Chapter 9

Many Questions

Introduction

Fallacies due to "making many questions into one" (henceforth, "Many Questions") arise

> when someone overlooks there being many [questions] and a single answer is given as if there is one question.[1]

Because questions in dialectic become the premises of arguments, this fallacy arises when one concedes a premise having multiple parts as though it were a simple proposition. The problem, then, amounts to an inability to individuate premises. As such, these arguments fail to be true refutations, because they violate clause (2) of the definition of a refutation: the conclusion does not follow necessarily from the premises. The reason for this is that what appears to be a premise is actually not one.[2]

A single premise is one that predicates "one thing of one thing."[3] In three different chapters of *de Interpretatione* (5, 8, 11), Aristotle amplifies this claim. Unfortunately, his comments about individuating statements in *de Int.* 5 appear to contradict his examples of the fallacy of Many Questions in *S.E.*, and his comments in *de Int.* 8 and 11 involve complications absent from any of the *S.E.* examples. In short, the *S.E.* treatment of the fallacy is much easier to deal with on its own, without bringing in the difficulties raised in the other treatise. Yet one cannot entirely ignore the *de Int.* material, for Aristotle does refer the reader at *de Int.* 11, 20b26, to "what was said about these things in the *Topics.*" This is best understood as a reference to his comments on the fallacy of Many Questions in *S.E.*[4] One must be careful, however, to isolate the material relevant to the unity of a premise from another issue that Aristotle raises in *de Int.* and elsewhere about the unity of a subject of definition. This second issue, I claim, is not at work in the fallacy of *S.E.* Accordingly, I will briefly look at *de Int.* only after I have examined the examples of the fallacy in *S.E.*

In what follows I argue the following points. Failures to predicate "one thing of one thing" arise in two general ways: either when the premise asked for involves a disjunction of terms, or when the premise asked for involves a conjunction of terms. In each case, there are two possible situations: a premise in which each of the terms of the disjunction or conjunction, when separated into single predications, has the same truth value as the other, or a premise in which the separated predications have differing truth values. These four possible situations are classified in the following chart:

(A) Disjunctive Premises "Is a or b (an) F?"	**(B) Conjunctive Premises** "Is a and b (an) F?"
(A1) a is F, and b is F a is not F, and b is not F	**(B1)** a is F, and b is F a is not F, and b is not F
(A2) a is F, and b is not F a is not F, and b is F	**(B2)** a is F, and b is not F a is not F, and b is F

Aristotle offers examples of three of these four possibilities (A1, B1, and B2). He believes that conjunctive premises are the more usual occasions of the fallacy of Many Questions, and he devotes most of his time to those examples. His position is that if one accepts a conjunctive premise whose single predications have differing truth values (B2), then the ensuing false refutation will arise because of a false presupposition that what is truly predicable of a class of things also is truly predicable of each member of that class, and vice versa. In other words, extralinguistic parts-wholes confusions lead one to accept the false refutation as true. On the other hand, if one accepts a conjunctive premise whose single predications all have the same truth value (B1), although there will not arise any apparent refutation, there will be no reasoning, because questioner and answerer may not be intending the same premise to be explanatory of the same conclusion. Finally, I show that Aristotle considers linguistic fallacies due to homonymy and amphiboly to be symptoms of this ontological error of not specifying proper explanatory premises.

Disjunctive and Conjunctive Premises

Aristotle's examples of the fallacy of Many Questions fall into two types. The question asked (i.e., the premise conceded) may have either a disjunctive or a conjunctive term. His examples uniformly place this multiple term as the subject, so the premise ends up saying one thing of more than one thing. However, a multiple predicate that says more than one thing of one thing illustrates the phenomenon equally well.[5] The two types of question, then, whose answers might lead one into a false refutation, are:

1. Is a or b (an) F? (Does F belong to a or b?)
2. Is a and b (an) F? (Does F belong to a and b?)

Disjunctive Premises

In cases of disjunctive terms, the questioner presents his opponent with an option: "Is a (an) F, or is b (an) F?" If one knows that only one of the options is true, there would seem to be no danger of fallacy in answering truly either that "a is F" or that "b is F." If both options, however, are known to have the same truth-value, a single answer that preserves that truth-value does not thereby preserve the reasoning from fallacy. Consider Aristotle's example of a disjunctive premise, both of whose disjuncts make false claims. He offers it as a multiple premise that is easily detectable and not a real sophistic threat:

> Which is sea, the land or the sky? (= Is it the land or the sky that is sea?)[6]

The assumption behind the question is that one or the other of the disjuncts is true.[7] Such disjunctive questions fail to constitute single premises, because they do not predicate one thing of one thing. In this case it predicates one thing (being sea) of two things (the land and the sky).

Aristotle's advice to someone confronted by such a question is simply not to answer.[8] But why not offer the simple answer "no" (i.e., "neither a nor b is F")? Although the answerer would be speaking truly, he would thereby be conceding the unity of the question. "For a single question is that to which there is a single answer."[9] Even though a single answer preserves the truth-value of each separately disjoined claim, it renders it unclear whether both of the disjuncts are being called explanatory of the conclusion, or just one. In the latter case, the argument may commit the fallacy of Non-Cause As Cause. Suppose, for example, that one is asked:

"Isn't it the case that the land or the sky is the sea?"

One answers: "No." (= "Neither the land nor the sky is the sea.")

"Are not whales sea-living animals?" "Yes."
"Therefore, whales are not land-living animals." "Yes."

The argument, with its disjunctive premise unpacked, runs this way:

1. The earth is not the sea.
2. The sky is not the sea.
3. Whales are sea-living animals.

Therefore, whales are not land-living animals.

But this is fallacious reasoning, according to Aristotle, because it offers an irrelevant premise (2) as explanatory of the conclusion. For the same reason, we should not expect Aristotle to allow a single answer to a disjunction each of whose terms makes a true predication. This failure to distinguish multiple premises, even if all of them are separately true or separately false, is treated more fully by Aristotle in his discussion of conjunctive premises, to which I now turn.

Conjunctive Premises

Aristotle contrasts the obviousness of the double question embedded in the "land or sky" disjunction with two examples said not to be so obvious. Both examples involve conjunctive questions, and both posit cases where the predication is true of one of the conjuncts and false of the other.

> "Things of which some parts are good and some parts are not good, are they all good or all not good?" For whichever one might answer, it is possible that one might seem to produce, as it were, a refutation, or an apparent falsehood. For to say that one of the good things is not good or that one of the things that are not good is good is a falsehood.[10]

We may formalize this apparent refutation in the following way:

Given that (1) a is F, and
(2) b is not-F,

the faulty premise-question is

(3) "Is (a-and-b) F or not-F?"

If one answers "(a-and-b) is F," then there arises the contradiction

(4) b is F and not-F, by (2) and the following false assumption (*):
(*) if (a-and-b) is F, then a is F and b is F.

If one answers "(a-and-b) is not-F," then there arises the contradiction

(5) a is F and not-F, by (1) and the same false assumption (*).

The false assumption (*) that underlies the acceptance of the premise amounts to the error of thinking that what is true of a conjunction is true of each conjunct. Here we have the real locus of extralinguistic parts-whole fallacies in *S.E.* It is nonlinguistic because it assumes an error in ontology: a predication that belongs to a whole also belongs to each part of the whole.

Aristotle's other example has received two divergent interpretations through history. I believe he intends it to illustrate the same sort of false refutation as just exemplified, this time presupposing the converse of (*), which I designate as (**): if a is F, then (a-and-b) is F.

> Is this and is this a man? As a result, if someone should strike this and this he will strike a man, but not men.[11]

At first glance the conclusion seems less a refutation than a solecism, which is a legitimate secondary goal of eristic.[12] It looks like a case of first predicating something of a group of individuals consecutively and then applying the same *grammatical form* (that is, singular rather than plural) of the predicate to the collective. If the "error" is simply a violation of proper Greek grammatical form, why should this not be a fallacy due to language? Aristotle's point, however, is not about the grammatical form of singular or plural nouns but about predicating of a plurality what is true of single members of the plurality.

The difficulty of interpretation here is to know what is being referred to by "this and this" (οὗτος καὶ οὗτος). Is the questioner pointing first to one person and then to another?[13] If so, then the fallacy does seem to take on a distinctly grammatical tone. For the predicate ("being a *man*") is now true of both members of the composite, and only the noun termination needs changing for its proper predication to the plurality ("being *men*"). Or is the questioner pointing first to one person and then to something that is not a person?[14] There are, I believe, three reasons to favor this second interpretation.

First, Aristotle's choice of the indeterminate pronoun οὗτος appears intentional. It can apply to both persons and non-persons. There seems no reason to use it if the fallacy could be made clear by using the names of two persons.

Second, Aristotle follows up the example by spelling out the false assumption (**) and illustrating its application to another example that involves a composite term of which contrary predications hold (e.g., man and not-man) rather than the same predication (e.g., man). He illustrates how the assumption of (**) can turn an apparent refutation into a genuine one (although based on the false assumption).

> Sometimes, when certain assumptions have been added [to the argument] there might even arise a genuine refutation, for example, if someone should grant that in the same way one thing and many things are called white or naked or blind. For if something which naturally sees but is without sight is blind, then too [many] things which naturally see but are without sight are blind. So whenever one thing has [sight] and another thing does not have sight, both of them will be either seeing or blind, which is impossible.[15]

Aristotle's point here requires some clarification. He is saying that if we grant that whatever it is that makes one thing blind in the same way also makes many things blind, then when among the many things there is something not blind there will be a refutation, for the set will be both blind and not blind. This is intended to explain our example: if that which makes οὗτος (e.g., Socrates) a man also makes οὗτος καὶ οὗτος (e.g., Socrates and Brunellus the ass) a man, so too that which makes οὗτος (e.g., Brunellus the ass) not a man will make the composite not a man. The result is the contradiction: οὗτος καὶ οὗτος both is and is not a man.

The third and most decisive reason for preferring this reading of οὗτος καὶ οὗτος is that Aristotle elsewhere explicitly denies that giving a single answer to a multiple question each of whose single predications admits the same answer leads one into contradiction or apparent refutation.

> Sometimes the predicate belongs to both and sometimes to neither, with the result that although the question is not simple, *nothing happens* to befall those who answer simply. . . . On the one hand, when many things belong to one thing or one thing to many, by answering simply and making this mistake *no contradiction follows*.

> On the other hand, when something belongs to one thing and not to another, or many things are predicated of many [then a contradiction may follow].[16]

Aristotle considers a simple answer to a question such as "Is Socrates and is Plato a man?" to be a dialectical mistake, but not one that will lead to contradiction. It is a failure in reasoning, but not a false refutation. But the answer to the οὗτος καὶ οὗτος question does lead to contradiction and refutation.

For all of these reasons I understand the οὗτος καὶ οὗτος subject to refer to two things, one of which the predicate belongs to and the other it does not. These preliminaries lead me to understand the abbreviated false refutation of our text in the following way. When asked "Is Socrates and is Brunellus a man?" if one answers "yes," then one is committed to the claim that by striking both, one strikes "men."[17] But in fact by striking both one strikes only one man (and an ass), not men. So when Aristotle writes:

> Is this and is this a man? As a result, if someone should strike this and this he will strike a man, but not men.

the result cited (ὥστε) is not the conclusion to the fallacious argument but the fact that contradicts the conclusion to the argument. The argument runs like this:

1. (multiple question): "Is not X (some person) and Y (some non-person) a man?"

[2. (**): If a is F, then (a-and-b) is F.]

[3. X is a man.]

4. (by 2 and 3): "Yes. X and Y are men."
5. Someone strikes X and Y.
6. (by 4 and 5): Someone strikes men.

But 6 is contrary to what actually results when someone strikes X and Y. What actually results (ὥστε) is that someone strikes a man, not men.

Resolutions of Fallacies Due to Many Questions

The danger of a false refutation due to Many Questions arises when at least one but not all of the many questions can be truly affirmed or denied. In order to resolve such a fallacy, "one must immediately distinguish the questions at the beginning."[18] That will explain why the refutation is false, for the conclusion does not follow from the premises but only from what appear to be premises. But why did the false refutation appear to be true? What erroneous belief about the world gave the sophism that convincing appearance? The answer is (*) or (**): the belief that what is predicable of a composite also is predicable of each part of the composite, and vice versa.

As was the case with disjunctive premises, Aristotle is not concerned only about conjunctive premises whose conjoined predications have opposite truth-values. Even if it so happens that each simple predication embedded in a multiple question admits the same answer, he insists that the answerer distinguish among the questions. The worry is not that answering truly will lead one into a false refutation, but that the single answer will render impossible the activity of reasoning.

> For it is not the case that if one answers the truth, for that reason the question is single. For it is possible for it to be true to say simply "yes" or "no" to countless different questions being asked, nevertheless one must not answer with a single answer, for it destroys the activity of reasoning (τὸ διαλέγεσθαι).[19]

This "activity of reasoning" must be understood as encompassing all four types of συλλογισμός introduced in *S.E.* 2.[20] These are not refutations, for they do not conclude with a denial of a previously accepted premise. But they are false or apparent συλλογισμοί. This breakdown of reasoning usually is due to one of the predications embedded in the multiple predication being true, but *not* being a necessary part of the account of the conclusion.

The greatest threat to proper reasoning by the acceptance of a multiple premise is, then, that the unpacked argument will commit the fallacy of Non-Cause As Cause. As I showed in my earlier treatment of Aristotle's fallacies of Begging the Question and Non-Cause As Cause (chapter 6), logical entailment from true premises does not suffice for proper reasoning. The premises must be properly explanatory as well as true. The requirement to be able to individuate premises in a multiple premise even if each single predication has the same truth-value amounts to a requirement to individuate arguments. It is necessary to the activity of reasoning that true premises failing to be explanatory of the conclusion not be conceded. To accept a multiple premise because *one* of the predications therein is explanatory is insufficient. It would be similar to an answerer accepting an amphibolous premise because both significations of the premise are acceptable, even though only one of the significations is a relevant part of the explanation of the conclusion.

Homonymy and Amphiboly As Cases of Many Questions

The analogy drawn above between amphiboly and Many Questions is not my own invention. Aristotle relates fallacies of verbal ambiguity to fallacies of Many Questions, and the connection he draws between the two is much closer than mere analogy. This is the one place in *S.E.* where he is willing to breach the strict divide separating fallacies due to language and fallacies outside of language. He uses the language of dependency: one type of fallacy arises because of another type of fallacy. What makes this particular relation important for an understanding of Aristotle's taxonomy is that the direction of dependency is the opposite of that of most modern (and ancient) reductionists. Instead of reducing all fallacies to linguistic equivocation, Aristotle reduces the verbal fallacies due to homonymy and amphiboly to the extralinguistic fallacy of Many Questions.

> If one did not make two questions into one question, neither would the fallacy due to homonymy and amphiboly arise, but there would be either refutation or not [i.e., no merely apparent refutations]. For what difference is it to ask if Callias and Themistocles are musical, or if there were one name applying to things that are different? For if [the name] signifies more than one thing, he asked many questions. If then it is not right to expect to receive simply one answer to two questions, it is clear that in no case is it proper to answer homonymous [questions] simply, not even if it is a true answer to all the questions, as certain people deem it. For this is no different than if one asked "Are Coriscus and Callias at home or not at home?", either when they are both present or both not present. For in both cases the premisses are multiple.[21]

If a word or phrase signifies many things, predications involving that word or phrase entail as many different premises as there are different significations.

Aristotle offers three examples in *S.E.* 30 of fallacious arguments due to Many Questions to illustrate the way these false reasonings also can be resolved as cases of verbal ambiguity. The first is the same example seen earlier of a composite of good and bad things being both good and not good, bad and not bad.[22] The other two are quite similar, one concluding of a composite subject that it is both different from and the same as itself, and the other that the composite is both equal to and unequal to itself. A close look at one of these arguments is enough to make clear Aristotle's point.

> If each thing is the same as itself and different from another, then since they are not the same as other things but as themselves, and also different from themselves, the same things are different from and the same as themselves.[23]

The argument works only on the false ontological presupposition of (**). Because A is the same as A, then A is the same as (A-and-B). And because A is different from B, then A is different from (A-and-B). The conclusion is that A is the same as and different from (A-and-B), which Aristotle signifies by "themselves." Aristotle follows the examples with his important conclusion:

> These [apparent refutations] also fall into other resolutions. For "both" and "all" signify more than one thing. Therefore it does not follow that they affirm and deny the same thing, except in name.[24]

The multivocity of "both" and "all" consists in their ability to refer either to two or more things separately or to two or more things together. Because of this, one can equivocate between the two uses and say that "both a and b" (together) are different from "both a and b" (separately), but the same as "both a and b" (together). The result is that "both a and b" are different from and the same as themselves.[25] This is not ambiguity in the sense (discussed in chapter 2) that μανθάνειν is ambiguous in Greek between being introduced to new knowledge and later applying that knowledge. The ambiguity is between two ways of predicating (separately or together) rather than between two different predicates.[26] Whatever differences we might see between double meanings due to homonymy and amphiboly, and the double meaning of words such as "both" and "all," both kinds of double meaning are said to result in a violation of clause (3a) of the definition of a refutation: "It does not follow that they affirm and deny the same thing, except in name." Even in cases where an arguer uses a homonymous word, yet the argument is unaffected by the multiple uses, Aristotle demands that the uses be distinguished.

> For when things are said in many ways, even if it makes no difference to the argument, the thought is necessarily uncertain if someone should make use of a term which can be distinguished in many ways as if it were not distinguished. For it is unclear according to what nature the word happens to belong to it [i.e., to what it signifies].[27]

In the end, Aristotle is willing to accept two different resolutions of the same false arguments. As fallacies due to double meaning, they fail to deny the same thing predicated by the answerer. But as fallacies due to Many Questions, the conclusion fails to follow necessarily from the premises, because the same premises have not been agreed upon by the two dialecticians.

Unity of Predication versus Unity of Definition: The Problem of *de Interpretatione*

We have seen that fallacies due to homonymy and amphiboly are reducible to fallacies of Many Questions in the sense that the resolutions of the former, which require one to distinguish between multiple significations of words, amount to resolutions of the latter, which require one to distinguish between different premises. Multiple linguistic significations reflect multiple ontological states signified by the different premises. The error that makes reasoning from multiple undistinguished premises, all of which are true, appear harmless is the false belief that conclusions are properly explained by any premises that logically entail that conclusion. A different ontological error was isolated for fallacies of Many Questions where the different predications embedded in the multiple premise have different truth-values. In these cases, the false refutation appears sound because of the false assumption of (*) or (**). When one looks at the passages in *de Int.* that touch upon the unity of predication, a further ontological issue seems evident. In what follows, I discuss these *de Int.* texts and argue that they confuse two different issues and cannot be used to further illuminate the *S.E.* fallacy of Many Questions.

de Interpretatione 5

This text appears to directly contradict Aristotle's strictures in *S.E.* against the unity of disjunctive and conjunctive premises. Aristotle writes:

> A single statement-making sentence is either one that signifies one thing, or one that is single by means of a connective. There are more [than one statement-making sentence] if more things, and not one, are signified, or if there are no connectives.[28]

The tension between these two criteria of propositional individuation has been thoroughly discussed by Ackrill.[29] Aristotle has conflated an ontological test ("signifying one thing") with a strictly grammatical test (the presence of connectives), even though the two tests do not produce coextensive sets. In *S.E.*, the only reason conjunctive and disjunctive questions fail to be single is because the multiple subjects and/or predicates lack ontological unity. Likewise, in the rest of *de Int.*, the grammatical test is ignored. It is hard not to charge Aristotle with carelessness in this grammatical account of the unity of a proposition.[30]

de Interpretatione 8 and 11

In *de Int.* 8, Aristotle comes to the ontological nub of the issue at hand. Individuation of statements requires that someone predicate one thing of one thing, and Aristotle begins an inquiry into what constitutes "one thing" and what does not. First he rejects the grammatical criterion of unity as a necessary condition for unity of predication. Even if there is one word for a composite (e.g., "cloak" referring both to man and to horse), predications involving that one word are not single predications, "because no man is a horse."[31] This is all we are told in *de Int.* 8. We are not told what constitutes one thing, only given an uncontroversial example of what does not constitute one thing. We can conclude from the example, though, that a conjunctive premise need not exhibit a grammatical conjunction. It need only signify conjoined predicates that are not "one thing."

In *de Int.* 11, the grammatical criterion is rejected also as a sufficient condition for unity. To predicate of man that he is "an animal and two-footed and tame" is to say, despite the conjunctions, one thing of one thing. But to predicate of man that he is "white and walking" is to say many things of one thing.[32] It is with this claim that the issue of the unity of a statement becomes mixed up with another issue: the unity of a definition. In an argument, premises such as "Man is an animal," "Man is two-footed," or "Man is tame" are all single predications in the same way that "Man is white" or "Man is capable of walking" are single predications. But if "Man is a two-footed animal" is a single predication, why not "Man is a white walker"? They may differ in truth-value, but as single predications, it is hard to see where they differ. Unfortunately, Aristotle never explains why "white walker" fails to signify one thing. He cannot use the reason given in *de Int.* 8 for why "man-and-horse" is not one thing, because there are walkers that are white. Usually commentators note that in *de Int.* 5 Aristotle postpones "to a different inquiry" a discussion of why, for example, a two-footed land animal is one thing and not many.[33] It is true that the subject comes up again and again in the *Metaphysics*,[34] but those discussions seem to me not very helpful for the issue of multiple predication. For in the *Metaphysics*, Aristotle is concerned with the unity of a subject of definition, which requires much stricter criteria than the unity of a subject of predication. It is commonplace to reason with simple predications whose terms do not signify a definitional unity.[35] It seems, then, that the task of picking out natural kinds or universals, which is at the heart of Aristotle's discussions of the unity of definition, is largely irrelevant to that unity of terms which Aristotle needs to individuate predications.

If the *de Int.* texts are as unhelpful as I suggest, the ontological fault with answering a question such as "Is Socrates and is Plato a man?" with a simple "yes" is not a failure to pick out essential unities.[36] Nor is there some obvious ontological blunder such as (*) or (**) at work. We have to go back to Aristotle's earlier diagnosis: it destroys the activity of reasoning by failing to avoid the incorporation of true but irrelevant matter in the premises of a given argument.

Conclusion

The fallacy of Many Questions involves the conceding of a multiple premise (conjunctive or disjunctive) as though the premise predicated only one thing of one thing, without distinguishing between the different predications. Two situations can arise with conjunctive premises. In the one, if the separated predications have different truth-values, a single answer can lead to contradiction. The refutation fails to be sound, because it draws its conclusion from what only appear to be premises but are not. The convincingness of that appearance arises due to a mistaken belief about extralinguistic parts and wholes: that what is true of a part is true of any whole that includes that part, or the converse, that what is true of a whole is true of any part of that whole.

In the second situation, an apparent argument can arise from a multiple premise all of whose separate predications have the same truth-value. Although one may answer such a multiple premise simply, without fear of eventual refutation, such an answer must still be avoided, because the result is the "destruction of reasoning." By this, Aristotle means that the dialectical partners may no longer be working on one and the same argument, despite the accident of verbal identity. And for there to be a proper argument at all, there must be no extraneous concessions in the premises.

The same need to individuate premises that is part of the resolution of fallacies due to Many Questions also may be necessary to resolve certain linguistic fallacies. In cases of premises with multiple significations due to homonymy or amphiboly, one needs to distinguish between the multiple meanings even when all of them may be true, so as to enable the answerer to specify which state of affairs is relevant to an understanding of the conclusion. Conceding a multiple premise amounts to a disregard for *which* premised predication is part of the intended explanation of the conclusion, even if it happens that each of the multiple predications is true. As in cases of Begging the Question and Non-Cause As Cause, the danger in not distinguishing premises lies in failing to specify which of the multiple premises is properly explanatory of the conclusion.

If we add to this conclusion the earlier discovery that fallacies due to Form of the Expression require for their resolution a knowledge of the distinctions among Categories, then we see that the removal of perplexity engendered by any one of the three fallacy types of double meaning requires more than knowledge about the nature and function of language. It requires some very particular understanding of the world itself. In order to resolve Form of the Expression fallacies, one must understand the different Categories of things in the world. In order to resolve fallacies due to homonymy and amphiboly, one must be able to distinguish which among different states of affairs is intended to explain the conclusion and which actually does. In short, contrary to more modern attempts to reduce apparent ontological confusions to linguistic confusions, Aristotle is showing that even among his fallacies due to language, many of them require specific ontological clarifications if they are to be avoided.

Conclusion and Summary

We have seen that Aristotle classifies false reasoning according to the following schema:

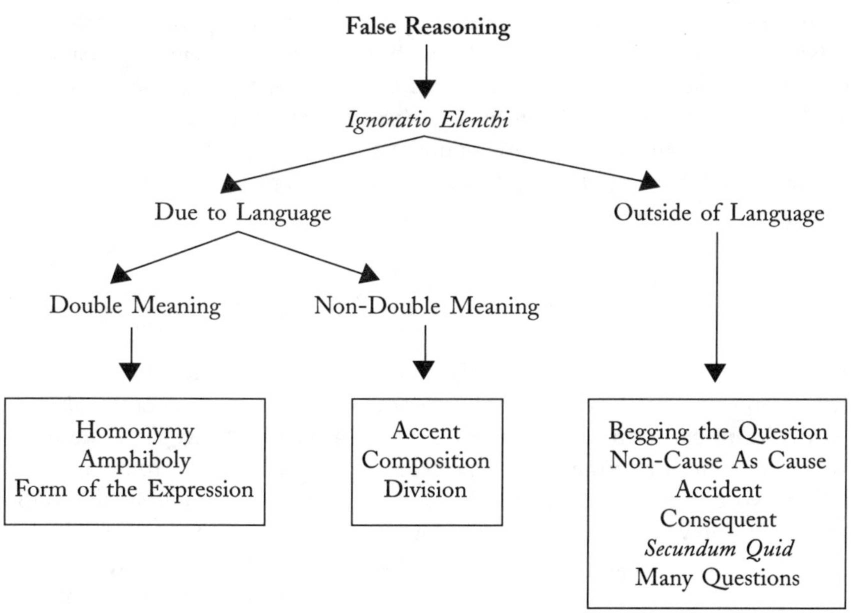

The first division into false reasonings due to language and false reasonings outside of language provided the impetus for this book. In order to discover the taxonomical criterion at work in that first division, however, it was necessary to understand each of the *infimae species* of false reasoning under those initial heads. I now summarize my findings and close with an amended schema that more accurately reflects Aristotle's practice of analyzing fallacies.

RESOLUTIONS

The criterion of division between linguistic and extralinguistic fallacies emerges out of Aristotle's notion of the resolution of a fallacy. Resolutions explain why a fallacy is a fallacy, thereby removing any perplexity arising from confrontation with the misreasoning. To fully understand a resolution, however, it is important to recognize that Aristotle employs a rich and an interchangeable vocabulary for fallacies, including both false (ψευδής) reasoning and apparent (φαινόμενος) reasoning. What makes an argument false and what makes an argument appear true are not the same, but both are essential to a resolution, because Aristotle does not distinguish between false arguments and apparent arguments. To explain why a piece of reasoning is *false* requires isolating the (logical) flaw that violates the definition of proper reasoning. To explain why the same piece of reasoning is *apparent* requires isolating that false presupposition that is accepted by the victim and would render the reasoning sound if it were true. The matter of these false presuppositions—that is, whether they concern the nature of language or the nature of the world—determines how Aristotle classifies that example of false reasoning. He places different false arguments under the same type of fallacy whenever the same false presupposition can account for the appearance of proper reasoning in each example.

THE PRINCIPLE OF PARSIMONY

It might be possible to assign highly unusual or idiosyncratic false presuppositions to different persons in order to account for their acceptance of each particular piece of apparent reasoning. Such a possibility would undermine any systematic understanding of fallacies by encouraging a proliferation of fallacy types to match the particular misapprehensions of innumerable individuals. To prevent such an explosion of fallacy types and to preserve fallacies as a legitimate domain for systematic study (τέχνη),[1] Aristotle applies another principle that limits the number of different fallacy types. I call this the "Principle of Parsimony." It is used to adjudicate among the possible different false presuppositions that might explain the appearance of proper reasoning. It opts for the most general of the possible errors, where generality is determined by how widely that particular presupposition can be satisfactorily applied to resolve other false arguments. So if a single false presupposition can account for the appearance of soundness in, say, ten different false arguments, whereas a different false presupposition can account for the appearance in only eight of them, Aristotle accepts the former as part of the correct resolution of the arguments.

False Presuppositions

The most general false presuppositions, which Aristotle believes are jointly necessary and sufficient to account for all examples of false reasoning, follow.

Errors about Language

1. Beliefs that linguistic signifiers (whether words, phrases, or entire propositions) are differentiated by their written transcriptions, regardless of their vocal pronunciation.
2. Beliefs that linguistic signifiers are differentiated by their component parts, regardless of the arrangement of those parts.
3. Beliefs that linguistic signifiers are always univocal.
4. Beliefs that linguistic units having similar morphology (or similar pronunciation) always signify similar kinds of entities.
5. Beliefs that linguistic units obeying similar syntactical rules always signify similar kinds of entities.

Errors outside of Language

1. Failures to recognize the differences among the Categories of being.
2. Failures to recognize the differences between essential and accidental predications.
3. Failures to recognize the differences between self-explanatory and non-self-explanatory things or states of affairs.
4. Failures to recognize which particular states of affairs possess special explanatory relationships to any particular non-self-explanatory state of affairs.

Criterion of Linguistic/Extralinguistic Division

Common to all of Aristotle's fallacies due to language is the presence of some false presupposition about language that produces the false appearance of proper reasoning. Common to all of Aristotle's fallacies outside of language, except for *Secundum Quid* fallacies, is the absence of any false presupposition about language to account for the false appearance of proper reasoning.

Number of *Infimae Species* of False Reasoning

Two of Aristotle's fallacy types cannot be justified as distinct kinds. Among the linguistic fallacies, Composition and Division are a single fallacy under two descriptions. Whichever description is used to identify the fallacy, the other identifies the opposite phenomenon necessary for resolving it. Among the extralinguistic fallacies, Aristotle fails to justify his examples of Consequent to be in any way different from his examples of Accident. In one telling instance, the same argument used to illustrate Accident is later repeated as an example of Consequent.

MULTIVOCITY OF "MULTIVOCITY"

Aristotle has a wider and a narrower version of linguistic multivocity. Under the wider version, any universal predicate is potentially ambiguous, because it can apply to multiple different individuals. He uses this version of multivocity in *S.E.* 1 to argue for the inevitable possibility of false reasoning due to language. When he begins his review of particular false arguments based on double meaning, however, his favored examples of homonymy and amphiboly involve words or phrases with semantic, not just referential, ambiguity. The distinction between homonymy and amphiboly is clear in these cases. Fallacies due to homonymy involve single names possessing semantic ambiguity independent of sentence context; fallacies due to amphiboly involve phrases possessing semantic ambiguity because of multiple syntactical roles for particular words in the phrase. When Aristotle tries to extend this distinction to arguments whose "multivocity" lies in universals applying to multiple possible individuals, his classification decisions become unclear.

This same confusion about what qualifies as linguistic multivocity intrudes upon the third type of double meaning fallacy: Form of the Expression. Although Aristotle cites distinctions between Categories as being necessary for resolving these errors, some of his examples involve confusions among different individuals of the same Category, or confusions between universals and particulars in the same Category.

In short, Aristotle would be better served by explicitly differentiating from semantic multivocity a second species of double meaning, namely, where one and the same signifier makes reference to multiple individuals under one universal signifier. When this multivocity results in fallacious reasoning, we might label this a linguistic fallacy of double meaning due to "Multiple Reference."

Taking into account all of the preceding conclusions, I offer on the following page a schema more accurately reflecting Aristotle's classification of false reasoning.

One final comment is in order. With the exception of the fallacies of Accent and Composition/Division, all of Aristotle's fallacy types derive their persuasive appearance in part or in whole from some sort of extralinguistic misconception. Because the ability to reason correctly is dependent on the

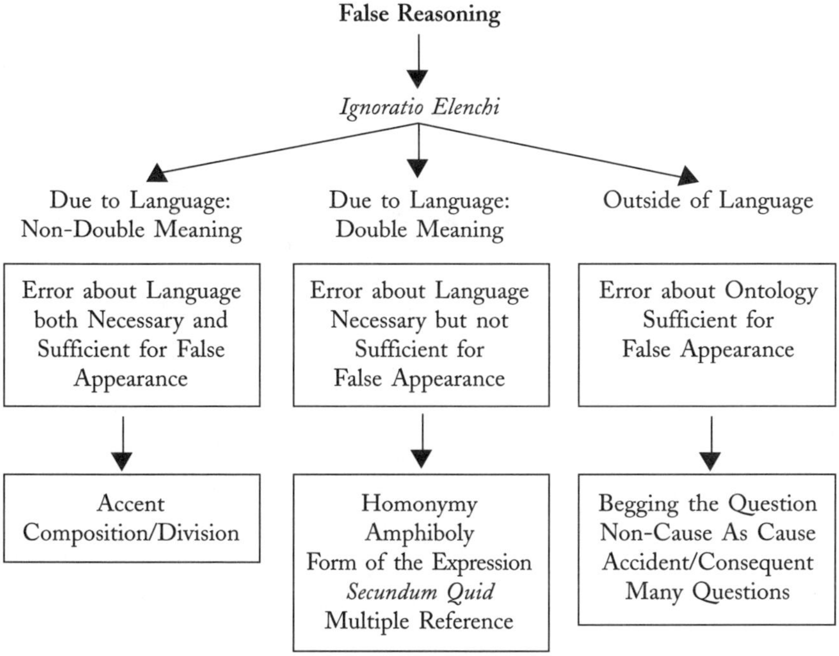

ability to distinguish true from false argumentation, it follows that, for Aristotle, proper reasoning requires a proper metaphysics. Nor is this just the claim that if one reasons from false premises, one might be led to false conclusions. Rather, it is to say that if one is mistaken in certain presuppositions about the world, one will not reason correctly even from true premises, because one will be unable to distinguish true from merely apparent arguments. What one believes affects one's ability to reason, not just the conclusions of one's reasoning. Although logic was considered in the peripatetic school that followed Aristotle propaedeutic, or preparatory, to philosophy, Aristotle at least did not think that one could preserve oneself from false reasoning without first being clear about certain substantive ontological claims, such as the order of explanation built into the universe, or the ontological distinctions among the real existents reflected by the Categories. It may not be that contemporary logicians must adhere to just those particular presuppositions that Aristotle isolates for proper reasoning. Aristotle has not given us the last word either on metaphysics or on language. But contemporary logicians should recognize, as Aristotle did, that even our logical intuitions—those beliefs about what constitutes a valid inference that formal systems of natural deduction are designed to capture—themselves are based upon definite ontological presuppositions. Aristotle's general position merits serious consideration. Logic, as the study of reasoning, is not a metaphysically neutral activity.

Appendix 1

Paralogisms in Aristotle

As I noted in the first note to chapter 1, Poste reads the καὶ at *S.E.* 1, 164a20–22,[1] as a true conjunction, according to which Aristotle distinguishes (1) sophistical refutations from (2) refutations that are only apparent and not real refutations, but fallacies (παραλογισμοί). Poste also tries to maintain this distinction between sophistical refutations and paralogisms in *S.E.* 8.[2] The most that I can read from *S.E.* 8, however, is that if the set of paralogisms is identical to the set of apparent-but-not-real refutations (Poste's reading of 164a20–22) then *if* one interprets those apparent-but-not-real refutations narrowly to include only invalid arguments (i.e., not to include valid arguments that only appear refutational) then paralogisms turn out to be a subset of the broader class of sophistical refutations.

Aristotle's terminology includes the following:

1. παραλογισμός (paralogism)
2. φαινόμενος ἔλεγχος/συλλογισμός (apparent argument)
3. ψευδὴς ἔλεγχος/συλλογισμός (false argument)
4. ἐριστικὸς ἔλεγχος/συλλογισμός (eristic argument)
5. σοφιστικὸς ἔλεγχος/συλλογισμός (sophistic argument)

Aristotle's texts reveal no consistent or even general distinctions in the use of these terms. The interchangeability of these labels may seem to run counter to expectations. One reason we might anticipate a distinction between (1) and (5) is the difference in how we would describe the victim of sophistic or eristic argument and the victim of paralogism. It is, we might say, the difference between being deceived and merely erring. Such a distinction was a part of the vocabulary of fallacy prior to Aristotle. Central to Plato's notion of "eristic" and "sophistical" was the desire to win an argument at any cost. This could, and often did, include the intent to deceive one's opponent. In Aristophanes, the depiction was even worse: intentional deception became the goal of sophistical argument. In the *Clouds*, the great attraction for Strepsiades in learning sophistic logic from the character Socrates was to be able to make the weaker argument appear to be the stronger, and thereby to outwit all of his creditors. Within this tradition, sophistic arguments are not the sort of mistakes someone commits privately. For Aristotle, however, the distinction between being tripped up by an unscrupulous dialectician and simply misreasoning in one's private researches is not preserved in the terminology of false reasoning. The effect in the mind of the victim is the same, whether that effect was intended or arose as a result of a private miscalculation. More importantly for Aristotle, the *cause* is the same. Whereas we are prone to emphasize the sophist as the cause of the deception in the former case, Aristotle restricts the causes of deception in both cases to one (or more) of the six sources παρὰ τὴν λέξιν or the six sources ἔξω τῆς λέξεως. The deception, whether abetted by another person or not, is effected by certain characteristics of language or by certain metaphysical confusions. In Aristotelian vocabulary, the sophist is, at most, only an efficient cause of the mistake in the victim's reasoning, much as a teacher is only an efficient cause of knowledge in the student. What "formally" accounts for the error is some particular linguistic or nonlinguistic confusion, not the sophist who might happen to encourage it. Our temptation to distinguish a paralogism from a sophistical argument by reference to the context of the error (in dialogue with a sophist or alone in one's study) does not conform to Aristotle's way of looking at the matter.

In *S.E.* 8, Aristotle does propose a distinction between paralogisms and sophistical refutations, although it is not Poste's distinction. However, it may have led Poste to his claim. Consider again Aristotle's terms:

1. παραλογισμός (paralogism)
2. φαινόμενος ἔλεγχος/συλλογισμός (apparent argument)
3. ψευδὴς ἔλεγχος/συλλογισμός (false argument)
4. ἐριστικὸς ἔλεγχος/συλλογισμός (eristic argument)
5. σοφστικὸς ἔλεγχος/συλλογισμός (sophistic argument)

What might have led Poste to his conclusion that Aristotle distinguished between (1) and (5) is the following analysis. Aristotle states at 169b20–23 that (2) is a subset of (5), since (5) includes not only apparent reasonings but also real reasonings that are only apparently relevant. Given that, plus the fact that Aristotle equates (1) and (2),[3] it follows that (1) is a subset of, rather than equivalent to, (5), which is the position that Poste argues. However, the claim that (2) is a subset of (5) at 169b20–23 is immediately preceded by Aristotle's introduction to the chapter in which he sets forth his purpose: to show that by knowing the sources of (2), we thereby also know the sources of (5) (69b18–19). This suggests that he is considering (5) to be a subset of (2)! What seems to be happening here is that Aristotle has both a narrower and a wider sense of φαινόμενος ἔλεγχος/συλλογισμός. In the narrower sense, it means just a formally fallacious argument (which, however, appears valid). In the broader sense, it includes all arguments that appear to accomplish something that they do not. This latter would include both apparently valid but formally invalid arguments and perfectly valid but only apparently relevant arguments.

Only in the narrow sense of (2) coupled with its equivalence to (1) can it be said that (1) is a subset of (5). Generally, though, Aristotle broadens φαινόμενος to cover any apparent reasoning, whether formally valid or not. As such, (1) and (5) are generally interchangeable. In an attempt to avoid this equation, Poste is compelled to emend the reference to παραλογισμοί at 170a9–11 to read either σοφιστικοὶ συλλογισμοί or ψευδεῖς συλλογισμοί. There is no textual support for this, and Poste seems motivated solely by the wish to keep paralogisms distinct from sophistical or false refutations.

As further evidence against the claim that Aristotle distinguishes sophistic or eristic argument from paralogism either according to the context of the error or intent of the perpetrator, consider the following examples cited by Aristotle, all of which involve efforts to deceive another and all of which are labeled "paralogisms."

1. *Poetics* 24, 1460a20: Aristotle cites Homer as one particularly skilled in "speaking falsehoods when necessary. This is a paralogism" (ψευδῆ λέγειν ὡς δεῖ. ἔστι δὲ τοῦτο παραλογισμός). The particular example cited is Homer's use of affirming the consequent (or conversion of a conditional).
2. *Rhetoric* II, 25, 1402b26: citing counterexamples to probability claims and claiming thereby to have refuted them (as though they were universal claims).
3. *Rhetoric* III, 13, 1414a6: the Homeric tactic of repeating a man's name several times in quick succession to give the impression of the character's importance.
4. Finally, in Aristotle's concluding chapter to *S.E.*, he summarizes his preceding work. "From how many and what kinds of sources fallacies arise in dialectic (γίνονται τοῖς διαλεγομένοις οἱ παραλογισμοί) . . . let these things we have said suffice concerning all [these matters]" (*S.E.* 34, 183a27–34). Here Aristotle describes *all* of the preceding false arguments, whether one is deceived by another or merely errs in good faith, as "paralogisms."

On the other hand, in *Physics* I, 2 (185a8), and I, 3 (186a6), the arguments of Parmenides and Melissus are called "eristic." Aristotle is not thereby attributing to those men bad will or the intent to deceive. I conclude that it is dangerous to attach too much significance to Aristotle's choice of one of his terms for fallacy over another, unless the context absolutely requires some distinction.

Appendix 2

Words and Counters—Platonic Antecedents

Aristotle's comparison (*S.E.* 1, 165a6–17) of the use of words in dialectic to the use of counters in arithmetic, and his concern with the vulnerability of the inexperienced (ἄπειροι) in both activities, is a reworking of a similar worry voiced by Adeimantus in *Republic* VI. There Socrates is arguing for the need for philosopher-rulers. He has just produced a long list of virtues that are agreed to characterize the philosophic nature. The climax of the argument is that precisely such philosophic virtues happen also to be the most desirable traits in a political ruler. Adeimantus sees the force of the argument but remains bothered by the apparent fact that the actual men who have devoted themselves to philosophy are either completely strange and unlike (ἀλλόκοτοι) Socrates' picture of the philosopher and totally depraved (παμπόνηροι), or, if not morally deficient, they are totally useless to the city because of their philosophic pastimes. In short, the conclusion of Socrates' argument runs contrary to actual Athenian experience with philosophers. And so Adeimantus voices this concern:

> Socrates, no one can dispute with you about these things. However, those who listen to you, every time they hear what you are now saying, they experience something like this; they think that because of their inexperience in the asking and answering of questions, they are being led astray by the speech, a little bit with each question, and when these little bits are gathered together at the end of the whole argument, a great fallacy appears, even a contradiction to what was said in the beginning. And just as those who are not well-practiced at checkers are held in check by those who are skillful, and they do not know what move to make, so also these who in the end are held in check, they too do not know what they should say, again because of this different kind of checkers game, one played not with counters but with words. When this happens, they do not believe that they hold the truth any better because of that experience. And I am speaking with an eye to the present case. For someone might now

> say that although by argument he does not know how to oppose you in each question, yet in matter of fact he sees that however many rush to pursue philosophy, and not those who, having engaged in it as young men so as to be educated, rid themselves of it, but those who dwelt upon it for a very long time, most of them are becoming altogether strange, not to say totally depraved, while the others who seem the most reasonable nevertheless suffer this from their pursuit of this thing which you praise: they are becoming useless to their cities.[1]

There follows Socrates' famous image of the ship of state with her trained pilot being ignored by the rest of the rowdy crew members. Although Aristotle has altered the second term of the analogy, from the counters used in a checkers game to counters used in arithmetical calculations,[2] the concern is the same: the failure to be convinced by an argument coupled with the inability to refute the argument.[3] Plato's task in *Republic* VI is to remedy Adeimantus' failure to be convinced by the argument; Aristotle's task in *S.E.* is to remedy his students' inability to refute sophistical arguments. Plato must explain why the facts adduced by Adeimantus only appear to conflict with the argument; Aristotle must explain why the argument only appears to conflict with an accepted fact.

Appendix 3

Aristotle on κύριον Predication

In *S.E.* 4 there appears a short enumeration of the causes of fallacies of double meaning.

> There are three ways of [arguing falsely] due to homonymy and amphiboly. [1] One is when either the phrase or the name standardly (κυρίως) signifies more than one thing, e.g., αἐτός and κύων. [2] One is when we are accustomed to speak in this way. [3] Third is when the combination [of words] signifies more than one thing, although in separation each signifies singly, for example, the combination "knowing letters" (ἐπίσταται γράμματα). For each word, both "knowing" and "letters," should it so happen, signifies one thing; but both [together] signify more than one, either the letters themselves have knowledge or another has knowledge of letters. So amphiboly and homonymy occur in these ways.[1]

To understand this short passage, it is necessary to briefly discuss Aristotle's doctrine of "standard" or κύριον predication. Many of the fallacious arguments due to language hinge upon the predication of names either "standardly" (κυρίως) or "nonstandardly." It is difficult to render κυρίως in English. Most translators render it in these contexts by "strictly" or "properly." I have chosen "standardly" (and the adjective "standard" for κύριον) rather than "properly" to avoid confusing Aristotle's κύριον ὄνομα with the English "proper name" in either its grammatical sense (where it contrasts with "common noun") or the various philosophical senses arising out of Russell's theory of descriptions (where grammatical proper names and definite descriptions are contrasted to logically proper names).

The difficulty of finding an English equivalent is due to the fact that Aristotle is attempting in his doctrine of κυρίον predication to respond to a debate between two extreme views of language. These two views are critically investigated in Plato's dialogue *Cratylus*. One view, the naturalist position, is defended by Cratylus. Names, according to this theory, are iconic. They imitate and make clear the essence of the thing named.[2] Because things have unique natures, there can be, by this theory, one and only one correct name for each thing "the same for all, both Greeks and barbarians."[3] In the dialogue, after developing the naturalist position, Socrates contrasts it to an alternative:

> the theory that Hermogenes and many others claim, that names are conventional and represent things to those who established the convention and knew the things beforehand, and that convention is the sole principle of correctness in names, and it makes no difference whether we accept the existing convention or adopt an opposite one according to which small would be called "great" and great "small."[4]

The disagreements between the naturalists and the conventionalists comprise a well-documented and long-standing dispute within the ancient world.[5] In the *Cratylus*, the spokesmen for the two positions—Hermogenes, the conventionalist, and Cratylus, the naturalist—appeal to Socrates for adjudication on the issue of the correctness of names. In typical fashion, Socrates first argues against Hermogenes in favor of a view of language naturally mirroring reality. But then he engages Cratylus to show the weaknesses of the naturalist position and the need for the role of convention in establishing correct names. Where these two Socratic critiques finally leave Plato is a matter of some debate.[6] What is clear is that Plato enlists his standard doctrine that the proper objects of knowledge are the extralinguistic Forms in order to defeat the extreme naturalism of Cratylus, for if names accurately make clear the essences (i.e., Forms) of things, then words themselves become legitimate objects of philosophic study. Etymology becomes a road to knowledge. Socrates' attack upon the naturalist position effectively defeats the belief that the study of names brings the sort of wisdom sought by the true philosopher. On the other hand, once Socrates has dethroned names from being the objects of study to being tools for studying the real objects, he acknowledges that, like all tools, some are better than others at their designated function.[7] So some names are better than others, and Socrates is forced back to the criterion of imitation for evaluating the relative utility of names. The dialogue ends with Socrates insisting that one must know the Forms apart from their names in order to properly judge the accuracy and correctness of those names.

The dialogue is aporetic, in that it leaves Socrates caught in a dilemma. The naturalist theory elevates names to objects of knowledge. This leaves etymology as the road to wisdom, replacing contemplation of the Forms as the preeminent philosophic task. Contrary to naturalism, words are mere tools. Different tools, however, are more or less suitable for accomplishing their assigned task. But strict conventionalism allows no room for better or worse names.

Aristotle is well aware of the difficulties Plato raises in *Cratylus* on both sides of the debate. He is keenly opposed to the naturalist position of Cratylus, but he is not unappreciative of the force of Plato's critique of pure conventionalism. The doctrine of κύριον predication is Aristotle's attempt to graft onto the conventionalism of Hermogenes an allowance for degrees of "correctness" in naming. For Aristotle, the names assigned to things, whether κύριον or not, have those assignments solely by convention.[8] These conventional assignments are always relative to particular language-speaking groups. Aristotle writes in the *Poetics*:

> Each name (ὄνομα) is either a standard word (κύριον), or a foreign word (γλῶττα), or a metaphor, or an ornamental word (κόσμος), or a made-up word (πεποιημένον), or an extended word (ἐπεκτεταμένον), or a contracted word (ὑφῃρημένον), or an altered word (ἐξηλλαγμένον). I mean by "standard" a word which particular groups (ἕκαστοι) make use of, and by "foreign" a word which other groups (ἕτεροι) make use of. As a result, it is clear that the same word can be both foreign and standard, but not to the same people. For to the Cypriots σίγυνον ("spear") is standard, while to us it is foreign.[9]

There follows a lengthy discussion of metaphor and brief treatments of made-up, extended, contracted, and altered words. The uses of these are licensed by Aristotle to poets for particular metrical effects. Standard and foreign words are mutually exclusive categories for any given group of speakers *at any given time*. This latter qualification is not explicit in Aristotle, but presumably he would accept that a one-time foreign word could become standard and a one-time standard word could lose its status as a language evolves. Σίγυνον may be a strange word for a spear among the Athenians, but if a Cypriot spear is found to have a distinctive shape, length, construction, and so on, σίγυνον may become κύριον among the Athenians for a spear constructed in the Cypriot manner. Or perhaps σίγυνον even could become entirely interchangeable with the standard Athenian word for spear so that one entity (a spear) possesses two standard names.

Given Aristotle's theory of conventionalism wherein the names used to signify something are established by local agreement, and given the power of those names to signify many things because of the indefinite numbers of particulars falling under each name, it will turn out that conventionalism leads, in time, to many names having a range of significations. We are left then with numbers of things, and often things of quite different kinds, customarily signified by the same name. The κύριον ὄνομα picks out the logically prior significate among the conventional range of significations. In short, the standard name is always among the conventional names of things, but not all conventional names for things are standard names.

The *S.E* 4 passage cited earlier is one of several texts that make this distinction clear. There (166a15–18) Aristotle cites three sources of double meaning fallacies.

> [1] One is when either the phrase or the name standardly (κυρίως) signifies more than one thing, e.g., αἐτός and κύων. [2] One is when we are accustomed to speak in this way. [3] Third is when the combination [of words] signifies more than one thing, although in separation each signifies singly.

The first source of amphiboly or homonymy is simply a word or phrase standardly signifying several things. Aristotle's examples are the homonyms αἐτός and κύων. The two words are standard names for an eagle and a dog. Both too are standard names for types of fish.[10] One type of false argument that Aristotle has in mind here involves the standard use of κύων to refer to the dog-star Sirius. In the *Rhetoric*[11] he cites as a fallacious argument due to homonymy someone who, wanting to praise a dog, offers as evidence some praise of the dog "in heaven." The same fallacy occurs when someone cites Pindar's praise of Pan whom, the poet says, "the Olympians call a dog of the great mother." It is doubtful that Aristotle considers κύων a standard name for Pan (at least not among the Athenians, if indeed, according to Pindar, among the Olympians)! We have here, I suggest, an example of the second source of homonymy and amphiboly, namely, when we are accustomed to apply a word to signify more than one thing, even when the application is nonstandard. It is not common use alone that determines standard application, since language-speaking groups can commonly apply a nonstandard word to something. Commonly used metaphors would seem to fall into this category, for example, Pan as "the dog of the great mother." For Aristotle, one of the major uses of metaphors is to fill in the gaps left by the lack of standard names for many extralinguistic things. For instance, the activity of "casting seed-corn" (τὸ τὸν καρπὸν . . . ἀφιέναι) is standardly signified by the word "sowing" (σπείρειν). But, says Aristotle, there is no given name (ὄνομα κείμενον = κύριον) for the phenomenon of flames being cast off from the sun, so this nameless activity is signified metaphorically as "sowing."[12] Presumably, then, were someone to argue for some claim about seed sowing by reference to the sun sowing its rays, his would be a sophistical argument due to homonymy caused by a word commonly (in this case, metaphorically), but not standardly, used to signify more than one thing. Nor is it only metaphors that have customary nonstandard applications. For example, the κύριον ὄνομα signifying the Athenians would have been 'Αθηναῖοι, but they were accustomed to say 'Ἀθῆναι as a nonstandard significans for the Athenians, as well as the standard significans for the city of Athens.

A good example of Aristotle's search for the standard name among the range of customary uses of a word occurs in *Topics* I, 7. The concern in that chapter is to distinguish the different ways "the same" (τ' αὐτόν) is used. Aristotle says first that we are accustomed (εἰώθαμεν) to apply the name to things that are the same in three ways: first, to things that are the same in number; second, to things that are the same in species; or third, to things that are the same in genus. Among these three customary uses, the first application (numerical sameness) has the most general agreement (μάλιστα δ' ὁμολογουμένος). But even the most generally agreed sense includes a range of applications. The "most standard and primary" (κυριώτατα . . . καὶ πρώτως) sense of numerical sameness is having the same name or definition; "the second" (δεύτερον) sense is having the same property; and "the third" (τρίτον) sense is having the same accident.[13] Aristotle's point is that among the conventional uses of "same," there are more and less *standard* applications. Whereas custom (εἰώθαμεν) and agreement (ὁμολογουμένος) determine the *range* of uses, Aristotle never suggests that convention plays any role in establishing which conventional signification is "the most standard."[14]

How then does Aristotle believe such distinctions among the conventional uses of a name are made? The answer, I believe, involves an appeal to the expert.[15] The clearest example of such an appeal comes from *Nicomachean Ethics* III, 4. The difficulty posed there is the determination of the correct object of desire (τὸ βουλητόν). Some say it is the good. But if that is so, then we are forced to say that in the not infrequent cases when people "desire" something that is not good (for them), it is not a real desire. On the other hand, some say that the object of desire is the apparent good. But because the apparent good varies between individuals, it will turn out that there will be no single thing that is *the* object of desire, only as many objects of desire as there are desirers. Aristotle's solution to the conundrum is to claim that both claims are correct, but neither one is the entire truth. The unqualified object of desire (τὸ ἁπλῶς βουλητόν) is the good, while the object of each person's desire (τὸ ἑκάστῳ βουλητόν) is the apparent good. This is a metaphysical distinction, not merely a linguistic one.[16] These two realities may or may not coincide. One characteristic of the good man (i.e., the moral expert) is that in him they do coincide. To discover then what the good is, Aristotle would presumably advise one to observe and learn from the particular desires of some recognized moral expert.[17] In the same way, says Aristotle, one determines what is really bitter or sweet or hot or cold by how they appear to the healthy man, for in the sick man what is really sweet and what is apparently sweet may not coincide. In general, Aristotle's analysis of faculties and their correlative objects allows for better and worse functioning of those faculties.

An analogous correlation exists between the human faculty of naming and the objects named. By the "naming faculty" we need only understand the human capacity for speech (i.e., intelligible communication). To be exercising the naming faculty, one need only be able to use conventionally agreed-upon descriptions. But Aristotle recognizes more or less useful exercises of such a faculty. For the philosophical understanding of the world, the application of a name κυρίως is the most effective exercise of that faculty. The measure and norm of that exercise is the expert in naming. Just as the κύριον object of desire is known by what the good man desires, so the κύριον signification of a name is made known by how the expert applies names. The expert is the one who can distinguish

> what kind of things are signified by the same and what by different kinds of expression (for a man who can do this is practically next door to the understanding of truth . . .).[18]

To say that one knows standard names by paying regard to the distinctions made by experts is to push back a step the real question. How does the expert make such distinctions? Aristotle never offers a general theory of how this is done. It is reasonable to suppose, though, that he considered himself such an expert, for he has left us numerous examples of the distinctions he made. *Metaphysics* Δ, in particular, stands as a monument to the energies that Aristotle expended on the task of distinguishing for a given name or phrase that which it signifies *most standardly* (κυριώτατα). It would require a detailed analysis of *Metaphysics* Δ to uncover the general principles used by the Aristotelian expert to distinguish, among the conventional applications of names, which are more or less standard.

In summary, when Aristotle claims that two different ways in which fallacies of double meaning arise are (1) when a name or phrase *standardly* signifies more than one thing, and (2) when we are *accustomed* to use a name or phrase to signify more than one thing, he is calling attention to the fact that not only is everyday conventional speech susceptible to linguistically generated false reasonings, but even the more precise κύριον speech of the expert is vulnerable. This is so for three reasons:

1. There exist entities with more than one standard name. For example, nouns such as ἱμάτιον and λώπιον are standard names for the same item (*Top.* I, 7; *Meta.* Γ, 4), as are verbs such as πορεύεσθαι and βαδίζειν (*Rhet.* III, 2, 1404b39–1405a1).
2. There exist words that are standard names for more than one entity (e.g., κύων and αἐτός).

3. Many things have no standard names at all. The need to signify them in order to bring them within the ambit of human intelligibility is a motive in the production of metaphor.

This absence of a simple isomorphism between standard names and things to be named multiplies the possibilities of false reasoning (whether intentional or inadvertent) due to homonymy and amphiboly.

To bring the discussion of *S.E.* 4, 166a14–23, to a close, the third and final cause of fallacious ambiguity deals with phrases whose component words are univocal, but in combination admit more than one reading. This is a simple restatement of what we have seen to be Aristotle's notion of amphiboly. In the example of "knowing letters," "letters" has a double syntactical role, either as a subject or an object of the verb. Thus the phrase is amphibolous.[19]

Appendix 4

Platonic and Academic Background to *Secundum Quid*

Aristotle's failure to remove *Secundum Quid* fallacies entirely from the linguistic sphere reflects an earlier use of ἁπλῶς, which had no ontological signification but marked solely a linguistic distinction. To see this, let us return to those problematic cases of things possessing contrary predicates in differently qualified ways, neither of which admits of unqualified belonging. In the case of an object half black and half white, false reasoning is encouraged by our belief that it must be correct to identify the object ἁπλῶς as one color or another. But this is precisely what Aristotle denies. It is wrong to believe that for any range of mutually exclusive predicates (e.g., colors), at least one of them must apply without qualification to any object that admits at least one member of that range in some way. Note how analogous this is to Aristotle's earlier linguistic observations about standard (κύριον) names (see Appendix 3). Just as something may have names, none of which is standard, so it may admit many true predicates, none of which belongs without qualification. As we noted above, Aristotle places ἁπλῶς and κυρίως on the same side of that line, the crossing of which results in fallacies of *Secundum Quid*. However, ἁπλῶς is a way of being, whereas κυρίως is a way of speaking. This distinction, however, originates with Aristotle, and he is following upon a tradition of philosophical usage that observed no such dichotomy.

If Plato had wished to mark Aristotle's distinction between being f without qualification (ἁπλῶς) and being f in some qualified way, his metaphysics of separated Forms distinct from sensible particulars partaking of both being and non-being would seem ideally suited for such a contrast. Certainly there are hints of a move in that direction. At *Meno* 73e, Meno has just finished claiming that justice is virtue. Socrates seeks clarification. "Is it virtue, or a particular virtue?" (πότερον ἀρετή . . . ἢ ἀρετή τις;) Meno is confused, and Socrates presents an analogy. "Take roundness about which I would say that it is a particular shape (σχῆμά τι) but not simply that it is shape (ἁπλῶς ὅτι σχῆμα)." This is the only occurrence of ἁπλῶς in the *Meno*, and although the theory of Forms is not explicit in the dialogue, the term would seem to provide Plato with a neat way to mark off the particulars that participate in a Form from the Form itself. But, in fact, Plato almost invariably will refer to the Form P using the intensive pronoun (e.g., σχῆμα αὐτό). Plato uses ἁπλῶς nearly always as a modifier of verbs of speaking.[1] In the passage cited above (*Meno* 73e 3–5), saying that roundness is a particular shape is contrasted not with saying that it is shape ἁπλῶς but with ἁπλῶς saying that it is shape. And Plato is generally consistent in his restricting ἁπλῶς to a manner of speaking, not a manner of being. For ontological purposes, Plato will always prefer αὐτός to ἁπλῶς.[2]

A good example of this difference between linguistic and ontological descriptions occurs in *Republic* IV. In this section of the dialogue, Socrates has shifted his attention from the justice of the city to the justice of the individual. He begins his analysis of the parts of the soul and their potential for conflict. At 437d, Socrates is discussing the appetitive part of the soul. He wants to distinguish "thirst" from thirst for a hot drink or for a cold drink or for much or for little or, in a word, for some particular kind of drink (437d 8–11). Glaucon agrees that thirst itself (αὐτὸ . . . τὸ διψῆν) is never a desire for anything other than for drink itself (αὐτοῦ πώματος) (437e 4–5). Socrates then likens the situation to knowledge. "Knowledge itself is knowledge of what is known itself. . . but a particular kind of knowledge is knowledge of a particular kind of thing."[3] He offers this example: When there arises knowledge of building a house the result is that it is called the art of housebuilding. The contrast here between P itself (αὐτό) and a particular P (τις) is an ontological distinction. That is, Plato is claiming that there is a difference between the state of "thirsting-for-a-drink" and the state of "thirsting-for-a-cold-drink." He does not want to claim, as we might, that these are two descriptions of a single state, the second being a more complete description than the first. Nor is Plato here appealing to his more fundamental ontological distinction between Forms and sensible particulars. "Thirst-for-a-cold-drink" is not a particular instantiation of, imitation of, or participation in the Form of Thirst Itself.[4] The ontological distinction between Forms and sensible particulars is one between two different realms of existence: those of Being and Becoming. Here, though, the distinction is between two distinct states, both within the realm of Becoming, for example, the difference between Glaucon's having a thirst and Glaucon's having a thirst for something cold. Socrates then cites this ontological difference to explain why we may not simply call (ἁπλῶς καλεῖσθαι) the knowledge of health and sickness "knowledge," but we call it "medical knowledge" by adding to it a certain quality (438e 4–8). Again, Plato only resorts to ἁπλῶς when making a linguistic observation. He reserves αὐτός for the ontological point.

Let us look at one more example. In the sole occurrence of ἁπλῶς in the *Protagoras*, Socrates is seeking the agreement of Protagoras, that there are no bad pleasures or good pains. Protagoras balks at such a claim. "I do not know, Socrates, if I should answer ἁπλῶς in this way, as you ask, that all pleasures are good and all pains are evil" (351c7-d2). The intrusion of ἁπλῶς occurs in conjunction with ἀποκριτέον. The issue for Protagoras is whether simply to agree that all S's are P's or to speak more circumspectly and allow for the possibility of some S that is not a P.

Plato's restricted use of ἁπλῶς—to modify verbs of speaking—was probably common in the Academy and thus a part of Aristotle's heritage. This becomes clear when we compare a passage from the *Meno* with Aristotle's explanation of the meaning of ἁπλῶς in the *Topics*. At *Meno* 78d, Socrates has obtained an account of virtue from his interlocutor.

> According to Meno . . . virtue is the acquisition of gold and silver. Do you add to this acquiring, Meno, the words 'justly' and 'piously,' or does it make no difference to you but even if one secures these things unjustly, you call it virtue none the less? (78d 1–6)

Meno objects that acquiring things unjustly cannot be called virtue. Although ἁπλῶς does not occur in the passage, it is a perfect Platonic context for the term. Had Meno claimed that even the unjust acquisition of gold and silver was virtuous, we can easily imagine Socrates responding: "So because it makes no difference how one comes to acquire gold and silver, you, Meno, would ἁπλῶς call virtue 'the acquisition of gold and silver.'" In this passage, the contrast with ἁπλῶς calling is "adding to this acquiring the words 'justly' and 'piously' (προστιθεὶς τούτῳ τῷ πόρῳ . . . τὸ δικαίως καὶ ὁσίως)." Aristotle appeals to this same linguistic criterion—the addition of words—to explain the meaning of ἁπλῶς in *Topics* II, 11.

> That which is [καλόν] ἁπλῶς is that which, *when nothing is added*, you will say is καλόν or the contrary. For example, you will not say that to sacrifice one's father is καλόν, rather that it is καλόν for some people. Therefore it is not ἁπλῶς καλόν. However you will say that to honor the gods is καλόν *without adding anything*, for it is ἁπλῶς καλόν. As a result, that which might seem to be καλόν or αἰσχρόν or something else of this sort *when nothing is added* will be called ἁπλῶς.[5]

The phrase "when nothing is added" (μηδενὸς προστιθεμένου) summarizes the Platonic and Academic use of the term. Yet already Aristotle has expanded ἁπλῶς to make ontological claims: things *are* and *are not* ἁπλῶς καλόν. In fact, this passage reveals an extraordinary linkage between the way we speak and the way things are. We may analyze the passage in this way:

(I) b29–30: Identification of the ontological and the linguistic.
That which is [καλόν] ἁπλῶς (ontological) is that which, when nothing is added, you will say is καλόν or the contrary (linguistic).

(II) b30–32: First argument: linguistic entails ontological.
For example, you will not say that to sacrifice one's father is καλόν, rather that it is καλόν for some people (linguistic). Therefore it is not ἁπλῶς καλόν (ontological).

(III) b32–33: Second argument: ontological entails linguistic.
However you will say that to honor the gods is καλόν without adding anything (linguistic), for it is ἁπλῶς καλόν (ontological).

(IV) b33–35: Conclusion restated.
As a result, that which, when nothing is added, might seem[6] to be καλόν or αἰσχρόν will be called ἁπλῶς.

What we find in Aristotle is a gradual takeover of the older linguistic use of ἁπλῶς by κυρίως, and the reserving of ἁπλῶς for a way of being f, which contrasts to being f in some respect.[7] What emerges in *S.E.* is that where a linguistic predicate can be applied κυρίως, the ontological predicate signified by the name will belong ἁπλῶς. Because the standard way of applying color predicates to persons is according to skin pigment, the Ethiopian is standardly called "dark." The ontological analogue to this is that darkness belongs to him without qualification. But the order of causation must not be confused. Darkness does not belong ἁπλῶς to the Ethiopian *because* of our convention of signifying his color κυρίως by the predicate "dark." Rather, the linguistic practice is derived from the recognition of the way colors belong to persons. Nor does this destroy the conventionality of the relationship between names and what they signify. Signifying by convention does not mean signifying arbitrarily or randomly.[8]

Notes

INTRODUCTION

1. In this book I render συλλογισμός either by its transliteration or by "reasoning." One must not restrict it to the three-term argument form of the *Prior Analytics*. I do not render the term by "deduction," because not all formally valid deductions constitute syllogisms in Aristotle's sense. Examples of these are examined in chapter 6.

2. *Top.* I, 1, 100a25–27: Ἔστι δὴ συλλογισμὸς λόγος ἐν ᾧ τεθέντων τινῶν ἕτερόν τι τῶν κειμένων ἐξ ἀνάγκης συμβαίνει διὰ τῶν κειμένων.

3. *Top.* I, 1, 100a27–29. *Pst. An.* I, 2, 71b16–25, details further restrictions on demonstrative premises. *S.E.* 2, 165b1–3, calls this sort of reasoning "instructive" (διδασκαλικός) because of its use in teaching.

4. *Top.* I, 1, 100a25–b23.

5. *Top.* 101a5–17. Aristotle, however, does not restrict the term παραλογισμός to nondialectical contexts. Elsewhere the term is interchangeable with any sort of false or sophistical reasoning. See Appendix 1 on the fluidity of Aristotle's terminology.

6. Peirastic is briefly defined at *S.E.* 2, 165b4–7, as a form of syllogism distinct from dialectic. Elsewhere, however, it is closely allied to dialectic. *S.E.* 11, 171b4–5, calls peirastic a kind of dialectical reasoning (διαλεκτική τις). *S.E.* 8, 169b23–25, calls peirastic a part of dialectic (μέρος τῆς διαλεκτικῆς). *Top.* VIII, 5, attempts to lay down some rules for peirastic reasoning within the context of dialectic. There Aristotle allows that the premise proposed may be "either commonly believed or commonly disbelieved . . . or neither" (ἤτοι ἔνδοξον ἢ ἄδοξον . . . ἢ μηδέτερον: 159a39), thereby distinguishing peirastic from proper dialectic. Peirastic is omitted from the *Top.* I, 1, classification of types of reasoning.

7. *Top.* I, 1, 100b23–101a4.

8. *S.E.* 2, 165b7–8: ἐριστικοὶ δ' οἱ ἐκ τῶν φαινομένων ἐνδόξων, μὴ ὄντων δέ, συλλογιστικοὶ ἢ φαινόμενοι συλλογιστικοί. Even when referring to eristic in the broader sense of false reasoning, Aristotle regularly describes the premises as endoxic (or apparently endoxic). He seems generally committed to the notion of eristic as parasitic upon dialectical rather than upon scientific reasoning. The detecting and resolving of false arguments and refutations particular to a given science Aristotle

assigns to the expert in that discipline (ὁ ἐπιστήμων); it is left to the dialectician to examine false reasoning when it arises from causes common to all sciences (*S.E.* 9). The causes of fallacy listed by Aristotle in *S.E.* are the κοιναὶ ἀρχαί of all false reasoning, whether demonstrative or dialectical.

9. *S.E.* 1, 165a21–23: ἔστι γὰρ ἡ σοφιστικὴ φαινομένη σοφία οὖσα δ' οὔ, καὶ ὁ σοφιστὴς χρηματιστὴς ἀπὸ φαινομένης σοφίας ἀλλ' οὐκ οὔσης.

10. *S.E.* 8, 169b22–23: φαινόμενον δὲ οἰκεῖον τοῦ πράγματος.

11. *S.E.* 16, 175a5–12.

12. Genetic accounts of Aristotle's logical works (e.g., Maier, Solmsen) distinguish *S.E.* as a later composition added to the rest of the *Topics*. Solmsen considers both *Topics* VIII and *S.E.* to date from Aristotle's Assos period. See J. L. Stocks (1933).

13. In *S.E.* 4 and 5, Aristotle cites *Ignoratio Elenchi* as one of the thirteen distinct sources, but in *S.E.* 6, he elevates it to the genus under which the other twelve fall.

14. Modern criticisms and proposed revisions of Aristotle's classifications can be found in Poste's notes to his translation of *S.E.* (1866), Kirwan (1979), and Hamblin (1970).

CHAPTER 1

1. *S.E.* 1, 164a20–22: Περὶ δὲ τῶν σοφιστικῶν ἐλέγχων καὶ τῶν φαινομένων μὲν ἐλέγχων, ὄντων δὲ παραλογισμῶν ἀλλ' οὐκ ἐλέγχων, λέγωμεν ἀρξάμενοι κατὰ φύσιν ἀπὸ τῶν πρώτων. I read the καὶ in line 20 as epexegetical, as do Pickard-Cambridge (Oxford) and Forster (Loeb). Poste reads the καὶ as a true conjunction, according to which Aristotle distinguishes (1) sophistical refutations from (2) refutations that are only apparent and not real refutations but fallacies (παραλογισμοί). See Appendix 1 for a detailed criticism of Poste's reading and for my general doubts about distinguishing paralogisms from sophistical or eristic arguments in Aristotle.

2. *S.E.* 1, 165a6–17: ἐπεὶ γὰρ οὐκ ἔστιν αὐτὰ τὰ πράγματα διαλέγεσθαι φέροντας, ἀλλὰ τοῖς ὀνόμασιν ἀντὶ τῶν πραγμάτων χρώμεθα ὡς συμβόλοις, τὸ συμβαῖνον ἐπὶ τῶν ὀνομάτων καὶ ἐπὶ τῶν πραγμάτων ἡγούμεθα συμβαίνειν, καθάπερ ἐπὶ τῶν ψήφων τοῖς λογιζομένοις. τὸ δ' οὐκ ἔστιν ὅμοιον· τὰ μὲν γὰρ ὀνόματα πεπέρανται καὶ τὸ τῶν λόγων πλῆθος, τὰ δὲ πράγματα τὸν ἀριθμὸν ἄπειρά ἐστιν. ἀναγκαῖον οὖν πλείω τὸν αὐτὸν λόγον καὶ τοὔνομα τὸ ἓν σημαίνειν. ὥσπερ οὖν κἀκεῖ οἱ μὴ δεινοὶ τὰς ψήφους φέρειν ὑπὸ τῶν ἐπιστημόνων παρακρούονται, τὸν αὐτὸν τρόπον καὶ ἐπὶ τῶν λόγων οἱ τῶν ὀνομάτων τῆς δυνάμεως ἄπειροι παραλογίζονται καὶ αὐτοὶ διαλεγόμενοι καὶ ἄλλων ἀκούοντες. For possible Platonic antecedents for this important analogy, see Appendix 2.

3. Diogenes Laertius, i, 59. See also Polybius, v. 26. 13.

4. In chapter 3 I analyze the Third Man Argument in terms of this "power of names."

5. *Phy.* III, 4–8.

6. Other references to individuals as ἄπειρα include: (1) *Top.* II, 2, which recommends that an investigation of a claim of universal predication should proceed by examining the subject class by species rather than by individuals (κατ' εἴδη καὶ μὴ ἐν τοῖς ἀπείροις) (109b14ff.); (2) *Rhet.* I, 2, which denies that individuals are ever the proper subject matter of art, because individuals are indefinite and unknowable (τὸ δὲ καθ' ἕχαστον ἄπειρον καὶ οὐκ ἐπιστητόν) (1356b32–3). For the connection between indefiniteness and unknowability see *Pst. An.* I, 24, 86a4ff., and the aporiai of *Meta.* B, 4, 999a27. See also *Meta.* α, 2, 994b22.

7. *de Int.* 2, 16a19–21: Ὄνομα μὲν οὖν ἐστὶ φωνὴ σημαντικὴ κατὰ συνθήκην ἄνευ χρόνου, ἧς μηδὲν μέρος ἐστὶ σημαντικὸν κεχωρισμένον.

8. See Irwin (1982).

9. *de Int.* 1, 16a16; *Pst. An.* II, 7, 92b7.

10. *de Int.* 2, 16a29–31; 3, 16b11–15.

11. *Cat.* 4, 1b25–27: Τῶν κατὰ μηδεμίαν συμπλοκὴν λεγομένων ἕκαστον ἤτοι οὐσίαν σημαίνει ἢ ποσὸν ἢ ποιὸν ἢ πρός τι ἢ ποὺ ἢ ποτὲ ἢ κεῖσθαι ἢ ἔχειν ἢ ποιεῖν ἢ πάσχειν.

12. There are in Greek many compound names that seem to be generated from simpler names. Aristotle recognizes these and cites as an example "pirate-ship" (ἐπακτροκέλης) (*de Int.* 2, 16a22–26). But he is unwilling to concede that the two seeming component simpler names are really names that can signify apart from the compound. See the discussions in Ackrill (1963), 115–17, and in Arens (1984), 42–43.

13. In general, Aristotle ignores the peculiarities of names of individuals. This is because individuals as such are not proper subjects of scientific discourse. Sophistical reasoning, however, is not limited to scientific investigations, and some of the eristic errors that Aristotle deals with are generated when names of individuals become interchanged with general descriptions in opaque contexts (see chapter 2). The failure to distinguish among types of names remains one of Aristotle's analytical deficiencies. It was recognized as such by the Stoics, who divided Aristotle's class of ὀνόματα into (1) names of individuals (e.g., "Socrates") called ὀνόματα, and (2) common or general names (e.g., "man," "horse") called προσηγορίαι. See Robins (1990), 32–33.

14. See e.g., *Phaedo* 78e2, 102b2; *Republic* 596a7; *Parmenides* 130e5. Aristotle assigns this position to Plato at *Meta.* A, 6, 987b7–10. The general concern in Plato is to show that universal predicates are derivatively applied to particulars from their primary referents, the Forms. But the extension of this derivation even to names of individuals is sometimes addressed (e.g., "Hermogenes" and "Socrates" in *Cratylus* 383). In general, though, Plato is no more interested in the names of individuals than Aristotle is.

15. The interpretation of Plato's position in the *Cratylus* on the issue of whether names are conventionally or naturally assigned is notoriously difficult. See Appendix

3 for discussion and references. Even though the position of Cratylus (that current names belong naturally to their referents) is refuted, there is an admission that the original names of things in some sense picked out the essential natures of those things (390e–391b).

16. A particularly strong echo of this legacy can be detected in *Meta.* Z, 8, where Aristotle notes that the name of a material composite often is derived from the name of the form or essence of that composite (1033b16–20).

17. *de Int.* 7, 17a38–40.

18. *G.C.* I, 6, 322b29–33: σχεδὸν μὲν οὖν, ὥσπερ καὶ τῶν ἄλλων ὀνομάτων ἕκαστον λέγεται πολλαχῶς, καὶ τὰ μὲν ὁμωνύμως τὰ δὲ θάτερα ἀπὸ τῶν ἑτέρων καὶ τῶν προτέρων, οὕτως ἔχει καὶ περὶ ἁφῆς.

19. In the Loeb edition, the force of this passage has been softened by a mistranslation: "Practically speaking, just as every other term *which* is used in several senses *is so used* owing to verbal coincidence or because the different senses are derived from different prior meanings, so it is also with 'contact' " (emphasis added). This follows Joachim's translation (1922), 141. In the Revised Oxford translation, the editor has revised Joachim's translation to read as I have translated it. Williams (1982), 21–22, also translates it as I have.

20. This claim for universal verbal ambiguity also is credited by Gellius to Chrysippus: "Chrysippus asserts that every word is by nature ambiguous, since two or more things may be understood from the same word." (Chrysippus ait omne verbum ambiguum natura esse, quoniam ex eodem duo vel plura accipi possunt.) See Gellius (1927), XI, 12. Unfortunately, we have little idea what exactly Chrysippus meant by this.

CHAPTER 2

1. Among the Alexandrian grammarians, for example, a λόγος was composed of λέξεις, where a λέξις was the minimal unit of signification. The ***TEXNH ΓΡΑΜΜΑΤΙΚΗ*** of Dionysius Thrax (c. 100 B.C.) defines λέξις as "the smallest part of a properly constructed sentence" (λέξις ἐστὶ μέρος ἐλάχιστον τοῦ κατὰ σύνταξιν λόγου). See Robins (1990), 39, and Kemp (1986), 350. For the Stoic distinction, see Pinborg (1975), 97–101. The only place where Aristotle limits the meaning of λέξις by contrast is at *Rhet.* III, 10, 1410b27–34, where he distinguishes the thought expressed (τὴν διάνοιαν τοῦ λεγομένου) from the language used to express it (τὴν λέξιν).

2. *Republic* III, 396c.

3. *Laws* VII, 816d.

4. *Apology* 17d.

5. *Laws* VII, 795e. This is a strange passage in which Plato says that there is one sort of dance that imitates the Μούσης λέξιν and another designed for physical fitness. He seems to be referring to a style of movement that somehow represents a style of poetic recitation.

6. *S.E.* 4, 165b27–30: τούτου δὲ πίστις ἥ τε διὰ τῆς ἐπαγωγῆς καὶ συλλογισμός, ἄν τε ληφθῇ τις ἄλλος καὶ ὅτι τοσαυταχῶς ἂν τοῖς αὐτοῖς ὀνόμασι καὶ λόγοις μὴ ταὐτὸ δηλώσαιμεν. I am treating δηλοῦν as interchangeable with σημαίνειν and translating it as "signify." Bolton (1976) argues that Aristotle contrasts the two verbs. Irwin (1982), 243–44, has replied in defense of their interchangeability.

7. Kirwan (1979).

8. *S.E.* 6, 168a23–28: τῶν μὲν γὰρ ἐν τῇ λέξει οἱ μέν εἰσι παρὰ τὸ διττόν, οἷον ἥ τε ὁμωνυμία καὶ ὁ λόγος καὶ ἡ ὁμοιοσχημοσύνη (σύνηθες γὰρ τὸ πάντα ὡς τόδε τι σημαίνειν), ἡ δὲ σύνθεσις καὶ διαίρεσις καὶ προσῳδία τῷ μὴ τὸν αὐτὸν εἶναι τὸν λόγον ἢ τὸ ὄνομα τὸ διαφέρον.

9. *Cat.* 1, 1a1–2: Ὁμώνυμα λέγεται ὧν ὄνομα μόνον κοινόν, ὁ δὲ κατὰ τοὔνομα λόγος τῆς οὐσίας ἕτερος.

10. See e.g., *G.C.* I, 6, 322b29ff. For other texts, see note 26 in chapter 5 and Hintikka (1973), 9.

11. To make the same point in the language of first-order logic: homonymy, as it applies to things, is a two-place predicate, whereas homonymy, as it applies to names, is a one-place predicate.

12. *Euthydemus* 275d–278b.

13. *S.E.* 4, 165b30–32: εἰσὶ δὲ παρὰ μὲν τὴν ὁμωνυμίαν οἱ τοιοίδε τῶν λόγων, οἷον ὅτι μανθάνουσιν οἱ ἐπιστάμενοι, τὰ γὰρ ἀποστοματιζόμενα μανθάνουσιν οἱ γραμματικοί.

14. *Rhet.* II, 24, 1401a29–31: καὶ τὸν τὰ στοιχεῖα ἐπιστάμενον ὅτι τὸ ἔπος οἶδεν· τὸ γὰρ ἔπος τὸ αὐτό ἐστιν. This is the reasoning forced upon Cleinias by Euthydemus in Plato's *Euthydemus,* 276e8–277b1.

15. *S.E.* 4, 165b32–34: τό τε ξυνίεναι χρώμενον τῇ ἐπιστήμῃ καὶ τὸ λαμβάνειν ἐπιστήμην.

16. *Euthydemus,* 277e5–278a5.

17. The strict conclusion to the false syllogism would be: ὁ καθήμενος ἕστηκεν. (The sitting person is standing.)

18. The strict conclusion to the false syllogism would be: ὁ κάμνων ὑγιαίνει. (The sick person is healthy.)

19. *S.E.* 19, 177a9–11: Τῶν μὲν οὖν παρὰ τὴν ὁμωνυμίαν καὶ τὴν ἀμφιβολίαν ἐλέγχων οἱ μὲν ἔχουσι τῶν ἐρωτημάτων τι πλείω σημαῖνον, οἱ δὲ τὸ συμπέρασμα πολλαχῶς λεγόμενον. Aristotle says "questions" because answers to questions are the premises of dialectical arguments.

20. *Top.* I, 10, 104a4–7: οὐ γὰρ πᾶσαν πρότασιν οὐδὲ πᾶν πρόβλημα διαλεκτικὸν θετέον· οὐδεὶς γὰρ ἂν προτείνειε νοῦν ἔχων τὸ μηδενὶ δοκοῦν οὐδὲ προβάλοι τὸ πᾶσι φανερὸν ἢ τοῖς πλείστοις. "For not every premiss nor every problem ought to be offered as dialectical. For no one in his right mind would put forward something that seems [true] to nobody, nor propose something that is obvious to everyone or to most people."

21. Hintikka (1973), 19–25; Owen (1986), 261.

22. Irwin (1981), 530.

23. Irwin also offers slightly different interpretations of three passages in the *Organon*: *Top.* 110b16–25; *Top.* 129b31–32, 130a9; and *S.E.* 166a15–16.

24. Hintikka (1973), 24, concludes that the difference between amphiboly and homonymy is insignificant for Aristotle. This is indicated, he writes, "by the fact that the word 'amphiboly' is often used by Aristotle for purposes other than the marking of the equivocity of a phrase. For instance, in *Poetics* 25, 1461a26, amphiboly is attributed to a single word." This claim that "amphiboly" *often* is used for other purposes is wrong. The example that Hintikka cites is the only indisputable application of "amphiboly" to what appears to be a single word. I argue below that this is only an appearance.

25. In English, it is word order that distinguishes "to wish me to capture the enemy" from "to wish the enemy to capture me."

26. *S.E.* 4, 166a8–9: καὶ γὰρ τὸν γινώσκοντα καὶ τὸ γινωσκόμενον ἐνδέχεται ὡς γινώσκοντα σημῆναι τούτῳ τῷ λόγῳ.

27. To a Greek speaker, the second premise and the conclusion make unambiguously different claims, chiefly because of the case of λίθος.

28. They are Homer's use of (1) γλῶττα, (2) μεταφορά, (3) προσῳδία, (4) διαίρεσις, (5) ἀμφιβολία, and (6) ἔθος.

29. *Poet.* 25, 1461a26: "παρῴχηκεν δὲ πλέω νύξ," τὸ γὰρ πλείω ἀμφίβολόν ἐστι.

30. How little of Homer it was necessary to actually quote in order to bring to the educated Greek's mind an entire passage or scene is exemplified in Plato's *Protagoras*. In Socrates' account of the gathered sophists and their admirers at the house of Callias, after describing the group of men surrounding Protagoras, he says: " 'And after him I recognized,' as Homer says, Hippias of Elis . . ." (315b9). The citing of the brief introductory clause τὸν δὲ μετ' εἰσενόησα from *Odyssey* XI, 601, is enough to recall the entire Homeric scene of Odysseus encountering the shades of the dead in Hades. The rhetorical effect of Plato's citation is to stress the contrast between Socrates and the sophists as similar to that between the living and effective Odysseus and the dead and futile shades.

31. Fragment 161 in Aristotle (1886).

32. *Iliad* 10, 252–3:

ἄστρα δὲ δὴ προβέβηκε, παρῴχηκεν δὲ πλέων νὺξ
τῶν δύο μοιράων, τριτάτη δ' ἔτι μοῖρα λέλειπται.

33. Lucas (1968), 244–45.

34. *Rhet.* III, 18, 1419a 20–21: ἀποκρίνασθαι δὲ δεῖ πρὸς μὲν τὰ ἀμφίβολα διαιροῦντα λόγῳ καὶ μὴ συντόμως.

35. *Top.* VIII, 7, 160a24–29.

36. This chapter of the *Topics* concludes with a statement that might seem to lend support to Irwin's position that Aristotle allows for amphibolous words. "When there

CHAPTER 3

1. Diodorus Siculus (1950) XII, 53: "[Gorgias] was the first to use the rather unusual and carefully devised forms of expression (τοῖς τῆς λέξεως σχηματισμοῖς), such as antithesis, sentences with equal members or balanced clauses or similar endings (ὁμοιοτελεύτοις), and other such devices."

2. When referring to one or more of those items listed in *Cat.* 4 or *Top.* I, 9, I shall use the upper-case "Categories."

3. *S.E.* 22, 178a4–6.

4. *S.E.* 22, 178a4–9: Δῆλον δὲ καὶ τοῖς παρὰ τὸ ὡσαύτως λέγεσθαι τὰ μὴ ταὐτὰ πῶς ἀπαντητέον, ἐπείπερ ἔχομεν τὰ γένη τῶν κατηγοριῶν. ὁ μὲν γὰρ ἔδωκεν ἐρωτηθεὶς μὴ ὑπάρχειν τι τούτων ὅσα τί ἐστι σημαίνει· ὁ δ' ἔδειξεν ὑπάρχον τι τῶν πρός τι ἢ ποσῶν, δοκούντων δὲ τί ἐστι σημαίνειν διὰ τὴν λέξιν.

5. I render Aristotle's technical phrase τόδε τι by "certain this" rather than "particular." Aristotle uses the phrase to refer to that which is ontologically and epistemologically primary among the things that are. In *Cat.* and *S.E.*, sensible particulars are the best candidates for that role. Later (*Meta.* Z), however, a better candidate is proposed that is not a sensible particular. This too is called a τόδε τι. I agree with Lear (1988) and Driscoll (1981), that τόδε τι has a consistent sense throughout Aristotle, although its extension changes in the course of his reflections on substance.

6. *S.E.* 7, 169a33–35: πᾶν τὸ κατηγορούμενόν τινος ὑπολαμβάνομεν τόδε τι, καὶ ὡς ἓν ὑπακούομεν.

7. *S.E.* 22, 178a33–36: ἐρωτήσας οὖν ὃ ἔχει, συνάγει ἐπὶ τοῦ ὅσα· τὰ γὰρ δέκα ποσά. εἰ οὖν ἤρετο ἐξ ἀρχῆς, [εἰ] "ὅσα τις μὴ ἔχει πρότερον ἔχων, ἆρά γε ἀποβέβληκε τοσαῦτα;" οὐδεὶς ἂν ἔδωκεν, ἀλλ' ἢ τοσαῦτα ἢ τούτων τι.

8. *S.E.* 22, 178b34–36: καὶ "ἆρ' ὅ τις οἶδεν, ἢ μαθὼν ἢ εὑρὼν οἶδεν; ὧν δὲ τὸ μὲν εὗρε τὸ δ' ἔμαθε, τὰ ἄμφω οὐδέτερον." ἢ ὃ μὲν ἅπαν, ἃ δ' οὐχ ἅπαντα.

9. *S.E.* 22, 178a39–b1: τὸ γὰρ μόνον οὐ τόδε σημαίνει οὐδὲ τοιόνδε οὐδὲ τοσόνδε, ἀλλ' ὡς ἔχει πρός τι, οἷον ὅτι οὐ μετ' ἄλλου.

10. *S.E.* 22, 178b8–9: "ἆρ' ᾗ Γονζᾶλεμὴ ἔχει χειρὶ τύπτοι ἄν;"

11. *S.E.* 22, 178b9–10: "ᾧ μὴ ἔχει ὀφθαλμῷ ἴδοι ἄν;" οὐ γὰρ ἔχει ἕνα μόνον.

12. *S.E.* 22, 178b11–13: οἱ δὲ ὡς καὶ ὃ ἔχει ἔλαβεν· ἐδίδου γὰρ μίαν μόνον οὗτος ψῆφον· "καὶ οὗτός γ' ἔχει," φασί, "μίαν μόνον παρὰ τούτου ψῆφον."

13. One ms. and Sophonias omit the τι. Such a reading places the adverb in the same syntactical position as a direct object, more readily encouraging the error.

14. *S.E.* 22, 178b5: τὸ γὰρ ταχέως οὐ τόδε διδόναι ἀλλ' ὧδε διδόναι ἐστίν.

15. *S.E.* 22, 178b6–7: ἡδέως ἔχων δοίη ἂν λυπηρῶς.

16. *S.E.* 22, 178b29–30: καὶ "ἆρ' ὃ μανθάνει ὁ μανθάνων, τοῦτ' ἔστιν ὃ μανθάνει; μανθάνει δέ τις τὸ βραδὺ ταχύ."

17. *S.E.* 22, 178b31: οὐ τοίνυν ὃ μανθάνει ἀλλ' ὡς μανθάνει εἴρηκεν.

18. *S.E.* 25, 180a 23–24.

19. In the *Euthydemus,* Plato recounts a particularly silly sophism that involves mistaking for a substance what has no Categorial significance whatsoever. If Aristotle were forced to classify the error, I suspect he would attribute it to Form of the Expression. In the dialogue, Ctesippus reacts to one of the sophistic displays of Euthydemus with the exclamation "Bravo Hercules!" (πυππὰξ 'Ηράκλεις). Dionysodorus, seizing upon (1) the similarity of the Greek vocative and nominative forms for names, and (2) the fact that the nonsignifying ejaculation πυππάξ is similarly positioned syntactically to be either grammatical subject or object, asks Ctesippus "whether you mean that Hercules is a bravo or that a bravo is Hercules?" (πότερον . . . ὁ 'Ηρακλῆς πυππάξ ἐστιν ἢ ὁ πυππὰξ 'Ηρακλῆς; 303a5–8).

20. *S.E.* 22, 178b24–27: "ἆρ' ὃ γέγραπται, γέγραφέ τις; γέγραπται δὲ νῦν ὅτι σὺ κάθησαι, ψευδὴς λόγος· ἦν δ' ἀληθής, ὅτ' ἐγράφετο· ἅμα ἄρα ἐγράφετο ψευδὴς καὶ ἀληθής."

21. *S.E.* 22, 178b27–29: τὸ γὰρ ψευδῆ ἢ ἀληθῆ λόγον ἢ δόξαν εἶναι οὐ τόδε ἀλλὰ τοιόνδε σημαίνει· ὁ γὰρ αὐτὸς λόγος καὶ ἐπὶ τῆς δόξης. This view that truth-values are qualities and that the same sentence (as substance) can change with respect to them agrees with Aristotle's position in *Cat.* 5, 4a22–b19, where he discusses the same example ("someone is sitting").

22. See also *de Int.* 4, 17a3–4, where being true and being false are said to "belong to" (ὑπάρχει) some sentences (affirmations and denials) but not to other sentences (e.g., requests).

23. *S.E.* 22, 178b32–33: ἢ οὐχ ὃ βαδίζει ἀλλ' ὅτε βαδίζει εἴρηκεν.

24. *S.E.* 22, 178a9–11: "ἆρ' ἐνδέχεται τὸ αὐτὸ ἅμα ποιεῖν τε καὶ πεποιηκέναι;" "οὔ." "ἀλλὰ μὴν ὁρᾶν γέ τι ἅμα καὶ ἑωρακέναι τὸ αὐτὸ καὶ κατὰ ταὐτὸ ἐνδέχεται."

25. *S.E.* 22, 178a16–19: εἰ δή τις ἐκεῖ, δοὺς μὴ ἐνδέχεσθαι ἅμα ταὐτὸ ποιεῖν καὶ πεποιηκέναι, τὸ ὁρᾶν καὶ ἑωρακέναι φαίη ἐγχωρεῖν, οὔπω ἐλήλεγκται, εἰ μὴ λέγοι τὸ ὁρᾶν ποιεῖν τι ἀλλὰ πάσχειν.

26. In *Meta.* Θ, 6, Aristotle explains the possibility of simultaneously seeing and having seen by distinguishing sight as an actuality (ἐνέργεια) rather than a motion (κίνησις).

27. *S.E.* 22, 178a11–16: "ἆρ' ἔστι τι τῶν πάσχειν ποιεῖν τι;" "οὔ." "οὐκοῦν τὸ τέμνεται καίεται αἰσθάνεται ὁμοίως λέγεται καὶ πάντα πάσχειν τι σημαίνει; πάλιν δὲ τὸ λέγειν τρέχειν ὁρᾶν ὁμοίως ἀλλήλοις λέγεται· ἀλλὰ μὴν τό γ' ὁρᾶν αἰσθάνεσθαί τί ἐστιν, ὥστε καὶ πάσχειν τι ἅμα καὶ ποιεῖν."

28. *S.E.* 166b10–19: Οἱ δὲ παρὰ τὸ σχῆμα τῆς λέξεως συμβαίνουσιν ὅταν τὸ μὴ ταὐτὸ ὡσαύτως ἑρμηνεύηται, οἷον . . . τὸ ποιὸν ποσὸν ἢ τὸ ποσὸν ποιόν, ἢ τὸ ποιοῦν πάσχον ἢ τὸ διακείμενον ποιοῦν, καὶ τἆλλα δ' ὡς διῄρηται πρότερον· ἔστι γὰρ τὸ μὴ τῶν ποιεῖν ὂν ὡς τῶν ποιεῖν τι τῇ λέξει σημαίνειν.

οἷον τὸ ὑγιαίνειν ὁμοίως τῷ σχήματι τῆς λέξεως λέγεται τῷ τέμνειν ἢ οἰκοδομεῖν· καίτοι τὸ μὲν ποιόν τι καὶ διακείμενόν πως δηλοῖ, τὸ δὲ ποιεῖν τι. τὸν αὐτὸν δὲ τρόπον καὶ ἐπὶ τῶν ἄλλων.

29. *S.E.* 22, 178b36–39: ὅτι ἔστι τις τρίτος ἄνθρωπος παρ' αὐτὸν καὶ τοὺς καθ' ἕκαστον· τὸ γὰρ ἄνθρωπος καὶ ἅπαν τὸ κοινὸν οὐ τόδε τι ἀλλὰ τοιόνδε τι ἢ ποσὸν ἢ πρός τι ἢ τῶν τοιούτων τι σημαίνει.

30. *Cat.* 5, 3b13–18: ἐπὶ δὲ τῶν δευτέρων οὐσιῶν φαίνεται μὲν ὁμοίως τῷ σχήματι τῆς προσηγορίας τόδε τι σημαίνειν, ὅταν εἴπῃ ἄνθρωπον ἢ ζῷον· οὐ μὴν ἀληθές γε, ἀλλὰ μᾶλλον ποιόν τι σημαίνει, —οὐ γὰρ ἕν ἐστι τὸ ὑποκείμενον ὥσπερ ἡ πρώτη οὐσία, ἀλλὰ κατὰ πολλῶν ὁ ἄνθρωπος λέγεται καὶ τὸ ζῷον.

31. Ackrill (1963), 88–89.

32. Both the Revised Oxford and Loeb translators render τοιόνδε τι as referring to the Category of Quality. This, I believe, is a mistake. Aristotle intends τοιόνδε τι as a contrast to particulars in any Category, not as a distinct Category itself. To adequately defend this point, I would need to offer a detailed reading of 178b39–179a10. There, I claim, Aristotle reserves ποιόν (179a6, 9) to signify the Category of Quality in contrast to τοιόνδε τι. I offer a brief account of my reading of this passage.

To help explain why people erroneously treat universals as particulars (in any Category), Aristotle cites the inseparability of the one from the many. He illustrates that confusion by comparing another inseparable kind (τοιόνδε) of particular with the particular itself (τόδε τι): musical Coriscus with Coriscus (178b39–179a3). Aristotle uses τοιόνδε at 179a2 to refer to the musical Coriscus, not to the Quality of musicality. In 179a3–8, Aristotle comments on the "musical Cortiscus" analogy: "Even if someone should say that what is [conceptually] separated [in the musical Coriscus example] is not what a particular is (ὅπερ τόδε τι) but what a Quality is (ὅπερ ποιόν) [i.e., musicality], it will make no difference. For there will still be some one besides the many, i.e., man." Aristotle's point is this: the Third Man Argument comes about by confusing universals and particulars. Inseparability explains why the confusion arises, just as in the case of Coriscus and the musical Coriscus. But if someone should say of this latter case that it is irrelevant to the Third Man problem because it confuses Quality with Substance (rather than universal with particular), there still remains the universal man besides Coriscus. I note, however, that the use of τοιόνδε τι for a universal in any Category is violated elsewhere by Aristotle. See, e.g., *S.E.* 22, 178a39–b1, where τοιόνδε signifies a different Category from τόδε, τοσόνδε, and so on.

33. *S.E.* 22, 179a8–10: φανερὸν οὖν ὅτι οὐ δοτέον τόδε τι εἶναι τὸ κοινῇ κατηγορούμενον ἐπὶ πᾶσιν, ἀλλ' ἤτοι ποιὸν ἢ πρός τι ἢ ποσὸν ἢ τῶν τοιούτων τι σημαίνειν. The suggestion here is that "man" does not signify a substance at all, which is hard to reconcile with the rest of Aristotle's writings on substance.

34. *S.E.* 22, 178b33–34: οὐδὲ τὸ τὴν κύλικα πίνειν ὃ πίνει ἀλλ' ἐξ οὗ.

35. *S.E.* 4, 166b11–12: οἷον τὸ ἄρρεν θῆλυ ἢ τὸ θῆλυ ἄρρεν ἢ τὸ μεταξὺ θάτερον τούτων.

36. The neuter definite article (τὸ) is commonly used by Aristotle when he wishes to refer to a word or phrase, but it cannot be regarded as either a necessary or sufficient syntactical sign for the distinction.

37. See Colson (1919) and Robins (1951, 1990) for accounts of this controversy. Grammarians were particularly interested in the correction of texts, especially Homeric texts. Where case endings of some word differed between texts, the analogists advocated the search for a similar nominative form whose oblique case endings would serve by analogy to determine the correct ending of the word in dispute. So, for example, if the genitive plural of θώς appeared as θώων in some texts and θωῶν in others, the analogists would cite the undisputed genitive ending θηρῶν of θήρ to argue for the latter reading. Analogists gave as their metaphysical justification for this procedure the orderliness of the world, which was to be reflected in the orderliness of language. Because this was chiefly a topic of debate among the Hellenistic schools, I do not deal with it here.

38. In ordinary Greek usage, the choice of definite article (ὁ or ἡ) would signal the sex of the fowl designated by ἀλεκτρυών with its masculine endings. In the case of "kneading-trough," the word ἡ κάρδοπος was a rare instance of a noun with masculine endings taking the feminine article. Yet it is clear that Aristophanes is doing more than poking fun at the grammatical anomaly. He proceeds to liken the nature of what the masculine word signifies to the (apparently well-known) effeminate nature of some man with the (masculine declension) name Κλεώνυμος (*Clouds*, 672–80).

39. *S.E.* 14, 173b19–22. See next section for discussion.

40. *Pol.* I, 5, 1254b13–16; I, 12, 1259a37–b10; I, 13, 1259b28ff.

41. *Meta.* Z, 5, 1030b18–26.

42. *Pst. An.* I, 4, 73a37–b1.

43. The "*per se* 2" label is from McKirahan (1992), 89. Wedin (1973) refers to these as "*per se* accidents." For difficulties with Wedin's treatment, see McKirahan (1992), 286, note 60.

44. *Rhet.* III, 5, 1407b6–9: τέταρτον, ὡς Πρωταγόρας τὰ γένη τῶν ὀνομάτων διῄρει, ἄρρενα καὶ θήλεα καὶ σκεύη· δεῖ γὰρ ἀποδιδόναι καὶ ταῦτα ὀρθῶς· "ἡ δ' ἐλθοῦσα καὶ διαλεχθεῖσα ᾤχετο."

45. *S.E.* 14, 173b17–22. I discuss this passage in the following pages.

46. Τὸ σκεῦος was used to refer to any sort of tool, implement, vessel, and so on. It was generally reserved for inanimate objects. Plato contrasts it to ζῷον, σῶμα, ψυχή (*Republic* X, 601d; *Gorgias* 506d). By the first century C.E., it could be applied negatively to persons to indicate their inferior status as tools, instruments, or chattels (Polybius), or positively to indicate their special status as divinely chosen vessels (New Testament). The important point is that σκεῦος was not used to refer to neuter word gender either by the later grammarians (e.g., Dionysius Thrax uses οὐδέτερα) or by Aristotle who uses μεταξύ in the grammatical discussion at *Poet.* 21, 1458a8–9. Kemp (1986), 344–45, credits the Stoics for introducing οὐδέτερον for the third gender in

place of the earlier σκεῦος and μεταξύ. At *S.E.* 14, 173b39–40, Aristotle distinguishes between males signified by names declined with masculine endings, females signified by names declined with feminine endings, and "what are called 'inanimate objects' possessing either masculine or feminine endings" (καὶ ἐπὶ τῶν λεγομένων μὲν σκευῶν, ἐχόντων δὲ θηλείας ἢ ἄρρενος κλῆσιν).

47. Cope (1867), 293–94.

48. *S.E.* 3, 165b20–21: τοῦτο δ' ἐστὶ τὸ ποιῆσαι τῇ λέξει βαρβαρίζειν ἐκ τοῦ λόγου τὸν ἀποκρινόμενον.

49. *S.E.* 3, 165b22: τελευταῖον δὲ τὸ πλεονάκις ταὐτὸ λέγειν.

50. *S.E.* 14, 173b17–22: ἔστι δὲ τοῦτο καὶ ποιεῖν καὶ μὴ ποιοῦντα φαίνεσθαι καὶ ποιοῦντα μὴ δοκεῖν, καθάπερ, ὃ Πρωταγόρας ἔλεγεν, εἰ ὁ μῆνις καὶ ὁ πήληξ ἄρρενά ἐστιν· ὁ μὲν γὰρ λέγων "οὐλομένην" σολοικίζει μὲν κατ' ἐκεῖνον, οὐ φαίνεται δὲ τοῖς ἄλλοις, ὁ δὲ "οὐλόμενον" φαίνεται μέν, ἀλλ' οὐ σολοικίζει. I retain Ross's reading of ὃ Πρωταγόρας in place of the mss. ὁ Πρωταγόρας. But I omit his quotation marks around ὁ μῆνις and ὁ πήληξ. Protagoras' point is not that the words are of a masculine declension, but that the things are of a male nature.

51. If Protagoras and his "man is the measure" doctrine has been accurately interpreted by Plato in the *Theaetetus*, it would be difficult for Protagoras to insist upon the distinction that Aristotle credits him with, that is, between an apparent and a real solecism. For according to Plato's account of Protagoras, whatever appears to someone to be the case *is* the case for that person. Despite the difficulty in squaring Plato's Protagoras with the position ascribed to him by Aristotle, even in the *Theaetetus* Protagoras is depicted as being thoroughly skeptical of the ordinary use of Greek language for the understanding of his philosophical position. See *Theaetetus* 168b6–c2.

52. *S.E.*, 14, 173b26–31: Εἰσὶ δὲ πάντες σχεδὸν οἱ φαινόμενοι σολοικισμοὶ παρὰ τόδε, [καὶ] ὅταν ἡ πτῶσις μήτε ἄρρεν μήτε θῆλυ δηλοῖ ἀλλὰ τὸ μεταξύ. τὸ μὲν γὰρ "οὗτος" ἄρρεν σημαίνει, τὸ δ' "αὕτη" θῆλυ· τὸ δὲ "τοῦτο" θέλει μὲν τὸ μεταξὺ σημαίνειν, πολλάκις δὲ σημαίνει κἀκείνων ἑκάτερον, οἷον "τί τοῦτο;" "Καλλιόπη, ξύλον, Κορίσκος." I follow Ross's reading of line 27. He omits the article before τόδε, and he brackets the following καὶ.

53. For multiple examples of such apparent solecisms, see Plato's *Protagoras* 330–331. See e.g., τοῦτο τὸ πρᾶγμα, ὃ ὠνομάσατε ἄρτι, ἡ δικαιοσύνη, αὐτὸ τοῦτο δίκαιόν ἐστιν ἢ ἄδικον; ("This thing justice which you just mentioned, is it not itself just or unjust?"), or again, σχολῇ μεντἄν τι ἄλλο ὅσιον εἴη, εἰ μὴ αὐτή γε ἡ ὁσιότης ὅσιον ἔσται ("Surely no other thing could be holy if holiness itself be not holy").

54. Another example that does not appeal to gender terminations may clarify the distinction that I am drawing between Protagoras and Aristotle. The Greeks sometimes used plural city names for a single city, for example, Ἀθῆναι for Athens, or Θῆβαι for Thebes. Aristotle would accept this conventional usage as an apparent-but-not-real solecism. One speculates that Protagoras would regard it as a real solecism that should be corrected by recasting the name into a singular form. For additional anomalies between word form and the kind of thing signified, see Robins (1957), 72ff.

55. There may, however, be occasion for the sophist himself to try to slip by his opponent a real but an unrecognized solecism. What sort of examples might Aristotle have in mind? If τοῦτο ἐστι Καλλιόπη is an example of an apparent-but-not-real solecism, we may suppose that αὕτη ἐστι Καλλιόπη, when it is a response to a query such as τί τοῦτο;, might be a real-but-not-apparent solecism. It is necessary to stress that only the particular dialectical context turns this into a solecism. In itself, the sentence is perfectly fine Greek, and to a question such as τίς αὕτη; it is the proper response. There also is a second way for Aristotle to allow for real-but-not-apparent solecisms. It too is dependent upon the particular dialectical situation. A real solecism may escape the notice of an insufficiently trained Greek speaker. There is no assumption that dialecticians are all equally habituated to proper Greek speech. Variations in linguistic competence can result in solecisms detected by some and undetected by others.

56. *S.E.* 14, 174a5–9: καὶ τρόπον τινὰ ὅμοιός ἐστιν ὁ σολοικισμὸς τοῖς "παρὰ τὸ τὰ μὴ ὅμοια ὁμοίως" λεγομένοις ἐλέγχοις. ὥσπερ γὰρ ἐκείνοις ἐπὶ τῶν πραγμάτων, τούτοις ἐπὶ τῶν ὀνομάτων συμπίπτει σολοικίζειν· ἄνθρωπος γὰρ καὶ λευκὸν καὶ πρᾶγμα καὶ ὄνομά ἐστιν.

57. Aristotle's distinction between solecism and errors due to Form of the Expression is, by my account, blurred by Protagoras. For Protagoras, solecism arises because words fail to reflect accurately the natures of the things they signify, even if the grammatical conventions are observed. For Aristotle, on the other hand, solecism is merely the violation of grammatical convention.

58. *S.E.* 7, 169a36–b2: διὸ καὶ τῶν παρὰ τὴν λέξιν οὗτος ὁ τρόπος θετέος, πρῶτον μὲν ὅτι μᾶλλον ἡ ἀπάτη γίνεται μετ' ἄλλων σκοπουμένοις ἢ καθ' αὑτούς (ἡ μὲν γὰρ μετ' ἄλλου σκέψις διὰ λόγων, ἡ δὲ καθ' αὑτὸν οὐχ ἧττον δι' αὐτοῦ τοῦ πράγματος)· εἶτα καὶ καθ' αὑτὸν ἀπατᾶσθαι συμβαίνει, ὅταν ἐπὶ τοῦ λόγου ποιῆται τὴν σκέψιν· ἔτι ἡ μὲν ἀπάτη ἐκ τῆς ὁμοιότητος, ἡ δ' ὁμοιότης ἐκ τῆς λέξεως.

59. The earlier *S.E.* 1 contrast between finite names and infinite things was between ὀνόματα and πράγματα.

60. See Ackrill (1963), 113.

61. *de Int.* 1, 16a3–9.

62. The presupposition of a conceptual gap between thought and speech is no Aristotelian, or even Greek, idiosyncrasy. It pervades Western philosophical reflection on language up until the nineteenth century. See Guttenplan (1974), 1–6.

63. Kirwan (1979), 42.

64. *S.E.* 7, 169a22–25: Ἡ δ' ἀπάτη γίνεται τῶν μὲν παρὰ τὴν ὁμωνυμίαν καὶ τὸν λόγον τῷ μὴ δύνασθαι διαιρεῖν τὸ πολλαχῶς λεγόμενον (ἔνια γὰρ οὐκ εὔπορον διελεῖν, οἷον τὸ ἓν καὶ τὸ ὂν καὶ τὸ ταὐτόν).

65. *Top.* I, 7, 103a6 ff.

66. *Meta.* Δ, 9 and I, 3.

67. *Meta.* I, 3, 1054a32ff.

Chapter 4

1. Stocks (1933) makes some use of the *Rhetoric* passages in his analysis relating *Topics* and *S.E.* to the two *Analytics*. Hamblin (1970), 70–73, notes certain discrepancies between the two accounts and speculates that they are due to Aristotle's habit of constantly revising earlier works.

2. At *Rhet.* II, 24, 1401a1–2, Aristotle says that one of the τόποι of merely apparent enthymemes is ὁ παρὰ τὴν λέξιν. He then divides this group into parts. The first is παρὰ τὸ σχῆμα τῆς λέξεως (1401a7) which, however, has a different and broader meaning than the particular fallacy type given that name in *S.E.* For details of the difference, see chapter 3 and Pp. 72ff. in this chapter. The second part is παρὰ τὴν ὁμωνυμίαν (1401a13). Then Aristotle introduces Composition and Division: ἄλλος τὸ <τὸ> διῃρημένον συντιθέντα λέγειν ἢ τὸ συγκείμενον διαιροῦντα. It is unclear whether this is supposed to be another part of those false enthymemes παρὰ τὴν λέξιν, or another source of error distinct from those παρὰ τὴν λέξιν.

3. Hamblin (1970), 71–72, for example, uncritically quotes the mistranslation as evidence of the discrepancy between *S.E.* and *Rhetoric.*

4. *Rhet.* II, 24, 1401a25–26: ἄλλος τὸ <τὸ> διῃρημένον συντιθέντα λέγειν ἢ τὸ συγκείμενον διαιροῦντα.

5. Language alone may not entirely account for the error, but some misunderstanding about the language used must play a necessary role.

6. The failure of many parts-whole confusions to be linguistically based is shown by Rowe (1962) and Cole (1965). Consider, for instance, this argument. "All the parts of this figure are triangular, therefore this figure is triangular." There is no problem in the language of this reasoning, either in signification of the chosen word or word arrangement. Rather it is the extralinguistic thought (i.e., the concept) of triangularity that brings about its failure to instantiate f in the formula, "If all the parts of x have f, then x has f." Some parts-whole confusions do seem to originate in λέξις. Consider this argument. "Each part of this machine is small, therefore the machine is small." One way to analyze it is to note that the relative (πρός τι) predicate "small" is being used with different senses in the premise and in the conclusion.

7. Although not a typical example of composition, Aristotle explicitly cites form and matter as "parts," because "they are the things into which the whole is divided or out of which the whole is composed, whether the whole is the form or that which has the form. For example, of a bronze sphere or a bronze cube, both the bronze is a part (this is the matter in which the form is) and the angle is a part" (*Meta.* Δ, 25, 1023b19–22).

8. *S.E.* 6, 168a23–28: τῶν μὲν γὰρ ἐν τῇ λέξει οἱ μέν εἰσι παρὰ τὸ διττόν, οἷον ἥ τε ὁμωνυμία καὶ ὁ λόγος καὶ ἡ ὁμοιοσχημοσύνη (σύνηθες γὰρ τὸ πάντα ὡς τόδε τι σημαίνειν), ἡ δὲ σύνθεσις καὶ διαίρεσις καὶ προσῳδία τῷ μὴ τὸν αὐτὸν εἶναι τὸν λόγον ἢ τὸ ὄνομα τὸ διαφέρον.

9. *S.E.* 20, 177a38–b2: ἔχει μὲν οὖν τι κἀκ τῶν ἀμφιβόλων ἐρωτημάτων, ἀλλ' ἔστι παρὰ σύνθεσιν. οὐ γάρ ἐστι διττὸν τὸ παρὰ τὴν διαίρεσιν. "It has

some [appearance] also of amphibolous questions, but it is really due to composition. For a fallacy due to division is not a double meaning fallacy."

10. *S.E.* 4, 166b1–3: Παρὰ δὲ τὴν προσῳδίαν ἐν μὲν τοῖς ἄνευ γραφῆς διαλεκτικοῖς οὐ ῥᾴδιον ποιῆσαι λόγον, ἐν δὲ τοῖς γεγραμμένοις καὶ ποιήμασι μᾶλλον.

11. *S.E.* 4, 166b3–6: οἷον καὶ τὸν Ὅμηρον ἔνιοι διορθοῦνται πρὸς τοὺς ἐλέγχοντας ὡς ἄτοπον εἰρηκότα "τὸ μὲν οὗ καταπύθεται ὄμβρῳ"· λύουσι γὰρ αὐτὸ τῇ προσῳδίᾳ, λέγοντες τὸ "ου" ὀξύτερον. The line in question comes from Homer's description of the chariot race in *Iliad* XXIII (328). Nestor is describing to his son the turning post around which the chariot race will be run. It consists of a dried tree stump flanked by two white stones, which was a monument set up long ago. On the "absurd" reading (οὗ), the line describes the monument as partially decayed. On the "corrected" reading (οὐ), the monument is not decayed. Presumably, such a long-standing monument would be inappropriately described as "partially decayed in the rain." Its longevity testified to its preservation from decay.

12. For a brief discussion of Aristotle's treatment of these Homeric texts, see Lucas (1968), 242–44.

13. *S.E.* 21, 177b37–178a2.

14. See also *de Int.* 1, 16a3–6: "written marks are symbols of spoken sounds."

15. *S.E.* 20, 177b4–7: ἀλλ' ἐν μὲν τοῖς γεγραμμένοις τὸ αὐτὸ <τὸ> ὄνομα, ὅταν ἐκ τῶν αὐτῶν στοιχείων γεγραμμένον ᾖ καὶ ὡσαύτως (κἀκεῖ δ' ἤδη παράσημα ποιοῦνται), τὰ δὲ φθεγγόμενα οὐ ταὐτά.

16. General use of such writing conventions only arrived a century later, when the Alexandrian scholars edited for (oral) reading earlier Greek poetry in non-Attic dialects, for example, Sappho.

17. If it is correct that fallacies of Accent, Composition, and Division were not effective in oral dialectic, then written representations of oral dialectical contests in which such fallacies occur cannot be historically accurate. Significantly, the version of Composition which appears in Plato's *Euthydemus* does not depend upon oral pronunciation. It is *not* an exclusively written fallacy. See pp. 68ff, "Confusing Linguistic Parts and Wholes" in this chapter.

18. As noted earlier and defended in the next chapter, I find no distinction between Composition and Division as fallacy types. In the *Rhetoric,* they are treated as a single fallacy. Subsequent references in this book will be to the single C/D fallacy.

19. *S.E.* 4, 166a23–30: παρὰ δὲ τὴν σύνθεσιν τὰ τοιάδε, οἷον τὸ δύνασθαι καθήμενον βαδίζειν καὶ μὴ γράφοντα γράφειν (οὐ γὰρ ταὐτὸ σημαίνει ἂν διελών τις εἴπῃ καὶ συνθεὶς ὡς δυνατὸν τὸ "καθήμενον βαδίζειν" [καὶ "μὴ γράφοντα γράφειν"]· καὶ τοῦθ' ὡσαύτως, ἄν τις συνθῇ τὸ "μὴ γράφοντα γράφειν"· σημαίνει γὰρ ὡς ἔχει δύναμιν τοῦ μὴ γράφων γράφειν· ἐὰν δὲ μὴ συνθῇ, ὅτι ἔχει δύναμιν, ὅτε οὐ γράφει, τοῦ γράφειν).

20. Kneale and Kneale (1962), 93.

21. Woods and Walton (1977), 382, also understand the Kneales to be ascribing these examples to homonymous equivocation. "Still, if the fallacy is an equivocation

on "possible," it is nonetheless a fallacy within language, since equivocation is a verbal fallacy." Woods and Walton are not opposed to this interpretation of Aristotle. Their concern is to keep the fallacy among those παρὰ τὴν λέξιν.

22. *S.E.* 20, 177b22–25: ἆρ' ὡς δύνασαι καὶ ἃ δύνασαι, οὕτως καὶ ταῦτα ποιήσαις ἄν; οὐ κιθαρίζων δ' ἔχεις δύναμιν τοῦ κιθαρίζειν· κιθαρίσαις ἂν ἄρα οὐ κιθαρίζων.

23. Aristotle says at *S.E.* 20, 177b25–26, it is not the case that someone who is not harping does not have the power of harping, but that he does not have the power of simultaneously doing it when he is not doing it (ἢ οὐ τούτου ἔχει τὴν δύναμιν, τοῦ οὐ κιθαρίζων κιθαρίζειν, ἀλλ', ὅτε οὐ ποιεῖ, τοῦ ποιεῖν).

24. *S.E.* 20, 177b29–31: οὐ γὰρ πάντως ὡς δύναται ποιεῖν δεδόσθαι ποιήσειν· οὐ ταὐτὸ δ' εἶναι ὡς δύναται καὶ πάντως ὡς δύναται ποιεῖν.

25. This principle, that it is characteristic of all examples of the same fallacy type to admit of the same resolution, is discussed in part 2.

26. *S.E.* 20, 177a36–38: ἆρ' ᾧ εἶδες σὺ τοῦτον τυπτόμενον, τούτῳ ἐτύπτετο οὗτος; καὶ ᾧ ἐτύπτετο, τούτῳ σὺ εἶδες;

27. *S.E.* 20, 177b10–12: οὐ γὰρ ταὐτὸ <τὸ> ἰδεῖν "τοῖς ὀφθαλμοῖς τυπτόμενον" καὶ τὸ φάναι "ἰδεῖν τοῖς ὀφθαλμοῖς" τυπτόμενον. The quotation marks are necessary additions by modern editors of the text. Without them, Aristotle is left saying that two identical signifiers (ἰδεῖν τοῖς ὀφθαλμοῖς τυπτόμενον) are not the same. The absurdity of this text apart from such modern devices to differentiate pronunciation is strong evidence that the error lies in confusing a transcription with a vocalized signifier.

28. Pickard-Cambridge (Revised Oxford) follows the suggestion of Uhlig to read the second word as ὀρός (i.e., the watery part of such fluids as milk, blood, semen, etc.). Whether the distinction between the two words is one of accent or breathing is irrelevant to Aristotle's point that the pronunciation differs, but the fourth-century B.C. written transcription does not.

29. τὰ φθεγγόμενα should not be thought of as being restricted to intentional language. They are not exclusively human-generated sounds. Animals, too, produce τὰ φθεγγόμενα. Even the booming of thunder, the creaking of doors, and the ringing of a pot when struck are all considered τὰ φθεγγόμενα. See Liddell and Scott (1940), 1927b.

30. *S.E.* 20, 177b4–7: ἀλλ' ἐν μὲν τοῖς γεγραμμένοις τὸ αὐτὸ <τὸ> ὄνομα, ὅταν ἐκ τῶν αὐτῶν στοιχείων γεγραμμένον ᾖ καὶ ὡσαύτως (κἀκεῖ δ' ἤδη παράσημα ποιοῦνται), τὰ δὲ φθεγγόμενα οὐ ταὐτά. ὥστ' οὐ διττὸν τὸ παρὰ διαίρεσιν.

31. *S.E.* 4, 166a36–37: ἐγώ σ' ἔθηκα <u>δοῦλον | ὄντ'</u> ἐλεύθερον. In the Greek text of this and the following examples of C/D, I have employed the devices of (1) underlining the words whose spoken combination results in one significant utterance, and (2) placing a vertical bar between words whose spoken separation results in a different significant utterance.

32. In this example and the next, Aristotle is citing Greek poetry where the possible locations of the vocal caesura were limited by the meter of the line. In this iambic trimeter, the natural poetic caesura would occur after ἔθηκα. But this leaves the nonsensical "free slave" combination (δοῦλον ὄντ' ἐλεύθερον) intact. The poet would need to place his caesura after δοῦλον to preserve the line from possible misunderstanding.

33. *S.E.* 4, 166a37–38: πεντήκοντ' ἀνδρῶν | ἑκατὸν λίπε δῖος 'Αχιλλεύς.

34. Again, the position of the caesura affects the significance of the line. In a dactylic hexameter, the caesura may occur after ἀνδρῶν or after ἑκατὸν. In the former position, we have the absurdity of Achilles leaving one hundred of the fifty men. In the latter position, πεντήκοντ' ἀνδρῶν ἑκατὸν is left undivided. See Adkins (1985), 1–5. Forster (Loeb) says that one possible reading of this line is "he left 150 men." The separation of πεντήκοντα from ἑκατὸν by ἀνδρῶν makes this a highly unlikely reading. Compound numerals are formed generally with καί: "150" = πεντήκοντα καὶ ἑκατόν (see Smyth 1920, 105). The position of ἀνδρῶν forces the reader or listener into one of the two aforementioned interpretations.

35. *S.E.* 4, 166a33–35: Παρὰ δὲ τὴν διαίρεσιν ὅτι τὰ πέντ' ἐστὶ δύο | καὶ τρία, καὶ περιττὰ καὶ ἄρτια, καὶ τὸ μεῖζον ἴσον· τοσοῦτον γὰρ καὶ ἔτι πρός.

36. Woods and Walton (1977), 382–83.

37. *Poet.* 20, 1456b39–1457a6: "A conjunction (σύνδεσμος) is a non-signifying sound which, when one signifying sound is composed out of several, neither prevents nor produces the combination, and which naturally stands both at the end and in the middle but must not be placed by itself at the beginning of the sentence, for example μέν or δέ. Or else it is a non-signifying sound which naturally makes one signifying sound out of more than one signifying sound."

38. *S.E.* 20, 177b13–15: ἆρ' ἔστιν ἀγαθὸν ὄντα σκυτέα μοχθηρὸν εἶναι; εἴη δ' ἄν τις ἀγαθὸς ὢν σκυτεὺς μοχθηρός· ὥστ' ἔσται ἀγαθὸς | σκυτεύς | μοχθηρός.

39. The linguistic source of this fallacy must not be confused with a different metaphysical issue that Aristotle raises in *de Int.* 11, using the same example of the "good cobbler." There Aristotle wonders why some kinds of predications that truly belong to things separately do not also truly belong to those same things when combined, whereas other kinds of predications truly belong both separately and combined. The predications in question are not linguistic units but extralinguistic entities. For instance, it is true that a man is two-footed and true that a man is an animal. It also is true of a man that he is a two-footed animal. Again, if it is true of something that it is a man and also true of it that it is white, then it is true that it is a white man. But in contrast to those two cases, Aristotle says: "But it is not the case that if something is a cobbler and it is good, then it is a good cobbler" (ἀλλ' οὐχί, εἰ σκυτεὺς καὶ ἀγαθὸς, καὶ σκυτεὺς ἀγαθός., *de Int.* 11, 20b35–36). The issue then is why some predicates that separately qualify their subjects, when joined to another predicate of that subject, transfer their primary attachment from the subject to that other predicate (e.g., the predicates signified by "good" or "evil"). On the other hand, some predicates, even when conjoined to other predicates, retain their original attach-

ment to the subject (e.g., the predicate signified by "white"). Aristotle thinks that this difference has something to do with the difference between accidental and essential predication, and whether the conjoined predicates produce a natural unity. This attempt in *de Int.* to find an ontological explanation for the different behavior of predicates is not successful (see Ackrill (1963), 146–48). But in *S.E.*, Aristotle is noting that in some cases where predicates transfer their primary attachment when joined to another predicate, a speaker can retain the primary attachment by his vocal pronunciation of the joint predication. In short, the issue here is not why some predicates exhibit this behavior and some do not; it is that spoken Greek can make clear whether or not that behavior is being signified.

40. *S.E.* 20, 177b16–20: ἆρ' ὧν αἱ ἐπιστῆμαι σπουδαῖαι, σπουδαῖα τὰ μαθήματα; τοῦ δὲ κακοῦ σπουδαία ἡ ἐπιστήμη· σπουδαῖον ἄρα μάθημα τὸ κακόν. ἀλλὰ μὴν καὶ κακὸν | καὶ μάθημα τὸ κακόν, ὥστε κακὸν μάθημα τὸ κακόν. ἀλλ' ἔστι κακῶν σπουδαία ἡ ἐπιστήμη.

41. *S.E.* 4, 166a31–32.

42. The Loeb translation interprets the sentence differently from mine. The two readings offered there are (A) "One single thing being able to carry, many things you can carry," and (B) "Being able to carry many things, you can carry one single thing only." The Revised Oxford translation also translates the sentence as (A) without offering a second reading. I find the reading of (B) to be possible but awkward. Apparently the vocalization of (B) would be τὸ ἓν μόνον δυνάμενον | φέρειν πολλὰ δύνασθαι | φέρειν. The only other commentator to try to explain the example is Sophonias (7, 27–29). He reduces the error to one of mistaking possibility with actuality: "Is it the case that he who can carry one thing is able also to carry many? Yes. Alcibiades is carrying one thing. Therefore Alcibiades, who is carrying only one thing, is carrying many" (τί δαί; ὁ ἕν δυνάμενος φέρειν δύναται καὶ πολλὰ φέρει; ναί. ὁ Ἀλκιβιάδης ἕν φέρει, ὁ Ἀλκιβιάδης ἄρα πολλὰ φέρει ἕν καὶ μόνον φέρων).

43. "[T]he phonetic features of ὑφέν included the loss or reduction of the pitch prominence of the accented syllable in one of the members of these compounds and close unions of two words. ὑποδιαστολή was clearly a phonetic mark associated with word divisions, particularly where without such a mark ambiguity would be possible" (Robins (1957), 80–83).

44. The error can arise at either the oral or the written level. In the former case, knowledge of all of the linguistic phonemes is falsely thought to entail knowledge of anything signified by the combination of those phonemes.

45. *Rhet.* II, 24, 1401a29–31: καὶ τὸν τὰ στοιχεῖα ἐπιστάμενον ὅτι τὸ ἔπος οἶδεν· τὸ γὰρ ἔπος τὸ αὐτό ἐστιν.

46. *Euthydemus* 276e–277b.

47. See the original (1928) Oxford edition.

48. *S.E.* 20, 177b12–13: καὶ ὁ Εὐθυδήμου δὲ λόγος "ἆρ' οἶδας σὺ νῦν οὔσας ἐν Πειραιεῖ τριήρεις ἐν Σικελίᾳ ὤν;". The question also might be translated, though

less naturally: "Don't you know now, being in the Piraeus, that there are triremes in Sicily?" The difference, I believe, is irrelevant to the fallacy.

49. See Freese's note, p. 327, to his translation of Aristotle's *Rhetoric* (Loeb) for an attempt at such an interpretation.

50. Cf. the "μανθάνει νῦν . . ." argument above.

51. *Rhet.* II, 24, 1401a28–29: ἔστι δὲ τοῦτο Εὐθυδήμου λόγος, οἷον τὸ εἰδέναι ὅτι τριήρης ἐμ Πειραεῖ ἐστίν· ἕκαστον γὰρ οἶδεν.

52. Grimaldi (1988), 342; Cope (1877), vol. 2, 307.

53. Cope (1877), vol. 2, 307.

54. *Rhet.* II, 24, 1401a31–34: καὶ ἐπεὶ τὸ δὶς τοσοῦτον νοσῶδες, μηδὲ τὸ ἕν φάναι ὑγιεινὸν εἶναι· ἄτοπον γὰρ εἰ τὰ δύο ἀγαθὰ ἕν κακόν ἐστιν. οὕτω μὲν οὖν ἐλεγκτικόν, ὧδε δὲ δεικτικόν· οὐ γάρ ἐστιν ἕν ἀγαθὸν δύο κακά.

55. *Rhet.* II, 24, 1401a34–36: πάλιν τὸ Πολυκράτους εἰς Θρασύβουλον, ὅτι τριάκοντα τυράννους κατέλυσε· συντίθησι γάρ.

56. *Rhet.* II, 24, 1401a36–b2: ἢ τὸ ἐν τῷ 'Ορέστῃ τῷ Θεοδέκτου· ἐκ διαιρέσεως γάρ ἐστιν· δίκαιόν ἐστιν, ἥτις ἂν κτείνῃ πόσιν, ἀποθνῄσκειν ταύτην, καὶ τῷ πατρί γε τιμωρεῖν τὸν υἱόν, οὐκοῦν καὶ ταῦτα πέπρακται· συντεθέντα γὰρ ἴσως οὐκέτι δίκαιον.

57. William of Sherwood (1966), 140.

58. Ibid., 141. The editor, Kretzmann, comments: "The substance of utterance (or of discourse) is the grammatically ordered string of words identifiable as a single expression. The acts of discourse known as Composition and Division are different ways of speaking (or reading) that string of words."

CHAPTER 5

1. *S.E.* 24, 179b17–18. The same point is illustrated at *Top.* VIII, 10, 160b23–39, which I discuss later in this chapter (Two Kinds of Resolution).

2. In the natural world, since nature "makes nothing in vain" (*PA* II, 13, 658a9), defective examples of natural kinds are themselves understood by discovering the necessary role they play in the preservation of the nondefective examples of the kind. So, for instance, Aristotle considers the production of female animal offspring to be a "departure from the type" (i.e., the type that is present potentially in the male semen). We understand the necessity of that departure, only because we know that without the division of animals into male and female, there could be no production of the nondefective type. See *GA*, IV, 3, 767a36–b13.

3. *S.E.* 8, 169b40–170a1.

4. *Meta.* A, 2, 982b12–15.

5. See, e.g., chapter 1 on signification.

6. *Top.* VI, 6, 145b16–20: ὁμοίως δὲ καὶ τῆς ἀπορίας δόξειεν ἂν ποιητικὸν εἶναι ἡ τῶν ἐναντίων ἰσότης λογισμῶν· ὅταν γὰρ ἐπ' ἀμφότερα λογιζομένοις ἡμῖν ὁμοίως ἅπαντα φαίνηται καθ' ἑκάτερον γίνεσθαι, ἀποροῦμεν ὁπότερον πράξωμεν. "Similarly, too, the equality of contrary arguments would seem to be productive of perplexity. For when it appears to us, upon reasoning both sides, that everything follows in the same way according to each conclusion, then we are perplexed over which of the two we should do." See also *Top.* I, 13, 104b12–14.

7. *S.E.* 8, 170a12–19.

8. *S.E.* 33, 182b6–12.

9. In only two passages does Aristotle grant that a certain type of false refutation may be resolved by reference *either* to a linguistic source *or* to an extralinguistic source: *S.E.* 17, 175b39–176a3, and *S.E.* 30. These are examined in chapter 9 (Many Questions).

10. τὸ μόνον τόδε. This is translated as "substance" in Loeb and "individual" in the Revised Oxford. On the former translation, Aristotle refers to the confusion of nonsubstances with substance. I note, however, that Aristotle writes τὸ τόδε, and *not* τὸ τόδε τι, which is the more characteristic formula for substance. (Ps.-Alexander, however, has τὸ τόδε τι.) On the other hand, if it refers to an individual, then the confusion is between universals and particulars. Both confusions, we saw in chapter 3, are encompassed within Form of the Expression.

11. *S.E.* 8, 170a12–19: Ἔστι δ' ὁ σοφιστικὸς ἔλεγχος οὐχ ἁπλῶς ἔλεγχος ἀλλὰ πρός τινα· καὶ ὁ συλλογισμὸς ὡσαύτως. ἂν μὲν γὰρ μὴ λάβῃ ὅ τε παρὰ τὸ ὁμώνυμον ἓν σημαίνειν καὶ ὁ παρὰ τὴν ὁμοιοσχημοσύνην τὸ μόνον τόδε, καὶ οἱ ἄλλοι ὡσαύτως, οὔτ' ἔλεγχοι οὔτε συλλογισμοὶ ἔσονται, οὔθ' ἁπλῶς οὔτε πρὸς τὸν ἐρωτώμενον. ἐὰν δὲ λάβωσι, πρὸς μὲν τὸν ἐρωτώμενον ἔσονται, ἁπλῶς δ' οὐκ ἔσονται· οὐ γὰρ ἓν σημαῖνον εἰλήφασιν ἀλλὰ φαινόμενον, καὶ παρὰ τοῦδε.

12. See *S.E.* 20, 177b31–34, for the contrast between resolutions πρὸς τὸν ἐρωτῶντα and those πρὸς τὸν λόγον. At *S.E.* 22, 178b16–17, Aristotle describes the same contrast between resolutions πρὸς τὸν ἄνθρωπον and those πρὸς τὸν λόγον. See also *S.E.* 33, 183a21–23.

13. Victims of false refutations can be anyone to whom the refutation appears sound. It may be the questioner, the answerer, or even third-party witnesses to the dialectical exchange.

14. *Top.* VIII, 10, 160b23–24: λυτέον ἀναιροῦντα παρ' ὃ γίνεται τὸ ψεῦδος.

15. *Top.* VIII, 10, 160b24–26: οὐ γὰρ ὁ ὁτιοῦν ἀνελὼν λέλυκεν, οὐδ' εἰ ψεῦδός ἐστι τὸ ἀναιρούμενον. ἔχοι γὰρ ἂν πλείω ψεύδη ὁ λόγος.

16. *Top.* VIII, 10, 160b33: οὐ γὰρ πᾶς ὁ καθήμενος γράφει.

17. See e.g., *S.E.* 20, 177b31–32: "of [false] arguments due to the same [fallacy], the resolution is the same" (τῶν γὰρ παρὰ ταὐτὸν λόγων ἡ αὐτὴ λύσις); *S.E.* 24, 179b11–12: "it is necessary that the correction of [false] arguments due to the same [fallacy] be the same" (δεῖ τῶν παρὰ ταὐτὸ λόγων τὴν αὐτὴν εἶναι διόρθωσιν).

18. In this respect, PP functions similarly to Aristotle's use of division in his zoological studies. That method picks out the essential natures of animals and issues in a final taxonomy. See Balme (1972), 105.

19. Aristotle addresses this criticism in *S.E.* 9, 170a20–34.

20. Cited in Hamblin (1970), 13.

21. See e.g., the comments of Wittgenstein in *Philosophical Investigations* I, 109, 132, 133.

22. See e.g., *S.E.* 10, 171a23–27: ἀλλ' οἱ μὲν πρὸς τοὔνομά εἰσι· καίτοι οὗτοι οὐ πάντες, οὐχ ὅτι οἱ ἔλεγχοι ἀλλ' οὐδ' οἱ φαινόμενοι ἔλεγχοι. εἰσὶ γὰρ καὶ μὴ παρὰ τὴν λέξιν φαινόμενοι ἔλεγχοι, οἷον οἱ παρὰ τὸ συμβεβηκὸς καὶ ἕτεροι. "Some arguments are relative to the word but these do not include all arguments, still less all apparent arguments. For there also exist apparent arguments not due to language, e.g., those due to accident, etc."

23. See chapter 9 on fallacies due to Many Questions.

24. Whately (1854), Book 3; Mill (1843).

25. *S.E.* 5, 167a21–27: Οἱ δὲ παρὰ τὸ μὴ διωρίσθαι τί ἐστι συλλογισμὸς ἢ τί ἔλεγχος [ἀλλὰ] παρὰ τὴν ἔλλειψιν γίνονται τοῦ λόγου· ἔλεγχος μὲν γάρ ἐστιν ἀντίφασις τοῦ αὐτοῦ καὶ ἑνός, μὴ ὀνόματος ἀλλὰ πράγματος, καὶ ὀνόματος μὴ συνωνύμου ἀλλὰ τοῦ αὐτοῦ, ἐκ τῶν δοθέντων ἐξ ἀνάγκης (μὴ συναριθμουμένου τοῦ ἐν ἀρχῇ), κατὰ ταὐτὸ καὶ πρὸς ταὐτὸ καὶ ὡσαύτως καὶ ἐν τῷ αὐτῷ χρόνῳ. The ἀλλὰ in the second line above is emended by Ross to ἄλλως. I follow the Revised Oxford edition and omit it entirely.

26. This captures the intent of Aristotle's καὶ ὀνόματος μὴ συνωνύμου ἀλλὰ τοῦ αὐτοῦ. The Revised Oxford translation omits ὀνόματος, despite the absence of any mss. or commentator support for such an omission. Presumably the emendation is inspired by the *Categories* doctrine, according to which synonymy and homonymy are relationships between things, not between words. But Aristotle also uses "synonymous" in our modern sense to signify a relationship between different words having the same sense and reference. See, for example: *Topics* VII, 13, 162b37. There are a number of ways in which dialecticians "beg the question" by assuming what needs to be demonstrated. Most of these are easily detectable. However, the mistake might be overlooked "in cases of synonyms, and in cases where the name and the phrase signify the same thing"; *Meta.* Γ, 4, 1006b18. "Signifying one thing" is not the same as "being said of one thing." For "white," "musical," and "man" are all said of the same thing, but they do not signify the same thing. If they did, then they would all be synonymous, and everything would be one; *Rhet.* III, 2, 1404b37–1405a2. Synonyms such as πορεύεσθαι and βαδίζειν are useful to the poet. Both are standard words for the same thing and synonymous with each other.

27. *S.E.* 4, 165b38–166a6.

28. *S.E.* 22, 178a26–27.

29. It is the importance of the common method of resolution that Kirwan overlooks when he questions the justification for treating Form of Expression as a type of double meaning. See chapter 3.

30. *S.E.* 6, 168a26–33: ἡ δὲ σύνθεσις καὶ διαίρεσις καὶ προσῳδία τῷ μὴ τὸν αὐτὸν εἶναι τὸν λόγον ἢ τὸ ὄνομα τὸ διαφέρον. ἔδει δὲ καὶ τοῦτο, καθάπερ καὶ τὸ πρᾶγμα ταὐτόν, εἰ μέλλει ἔλεγχος ἢ συλλογισμὸς ἔσεσθαι, οἷον εἰ λώπιον, μὴ ἱμάτιον συλλογίσασθαι ἀλλὰ λώπιον. ἀληθὲς μὲν γὰρ κἀκεῖνο, ἀλλ' οὐ συλλελόγισται, ἀλλ' ἔτι ἐρωτήματος δεῖ εἰ ταὐτὸν σημαίνει, πρὸς τὸν ζητοῦντα τὸ διὰ τί.

31. Aristotle's original description of clause (3b) (a denial also "of a name which is not synonymous but the same": καὶ ὀνόματος μὴ συνωνύμου ἀλλὰ τοῦ αὐτοῦ) is overly influenced by the "mantle" and "cloak" illustration. They are examples of synonymous but not identical names. Real examples of fallacies due to C/D or Accent involve names (or phrases) neither synonymous nor the same.

32. *S.E.* 23, 179a11–15: Ὅλως δ' ἐν τοῖς παρὰ τὴν λέξιν λόγοις ἀεὶ κατὰ τὸ ἀντικείμενον ἔσται ἡ λύσις ἢ παρ' ὅ ἐστιν ὁ λόγος. οἷον εἰ παρὰ σύνθεσιν ὁ λόγος, ἡ λύσις διελόντι, εἰ δὲ παρὰ διαίρεσιν, συνθέντι. πάλιν εἰ παρὰ προσῳδίαν ὀφεῖαν, ἡ βαρεῖα προσῳδία λύσις, εἰ δὲ παρὰ βαρεῖαν, ἡ ὀξεῖα.

33. William of Sherwood (thirteenth century) observed this point in his discussion of Accent: "I raise the following questions, however. Composition and Division are to an expression as Accent is to a word; therefore if there is more than one fallacy in the former case, so is there in this case [of Accent]. Alternatively, if there is [only] one in this case, so is there in the former case." William goes on to say that there is even less distinction between an act of Composition and one of Division than between two words differently accented (145–46).

34. *S.E.* 4, 166a23–30, discussed in chapter 4, pp. 60ff.

35. *S.E.* 20, 177b13–15, discussed in chapter 4, p. 66.

36. The same procedure of resolving by opposites is prescribed for fallacies of double meaning. If one is led to affirm that x is f after originally claiming that x is not-f, then one must resolve the apparent refutation by taking the "opposite name" (i.e., "not-f") and showing the sense in which it remains true of x (*S.E.* 23, 179a15–20). This amounts to an appeal to the "opposed" signification of the homonymous word or amphibolous phrase. In fallacies due to Form of the Expression, the appeal to opposites is a little more contrived. The "opposite" to which one appeals for resolution is the Category opposed to the one assumed by the questioner. Take, for instance, the argument that falsely refutes the claim that it is impossible to give away what one does not have. After conceding that someone with ten dice does not have only one die, and that the same person is able to give away only one die, the answerer must still deny that such a person can therefore give away what he does not have. Rather, he must appeal to the "opposite" Category that is being confused with substance and explain first that "only one" signifies the manner of giving something and second that the argument only shows that it is possible to give something away in a manner different from the manner in which one possessed it. Aristotle's rationale for considering fallacies due to Form of the Expression, like those due to homonymy and amphiboly, to be based on double meaning, is that they all are resolved by appeals to some signification opposed to that signification on which the apparent refutation rests.

37. Or if the ambiguity is in the conclusion, the need is to judge which state of affairs is adequately explained by the premises.

CHAPTER 6

1. Aristotle uses as labels for this fallacy τὸ ἐν ἀρχῇ αἰτεῖσθαι and τὸ ἐν ἀρχῇ λαμβάνειν. Literally these would be rendered as "asking the original point" and "assuming the original point." Because the English translation "Begging the Question" for the Latin label "petitio principii" is part of standard philosophical nomenclature, I retain this traditional label.

2. *Pr. An.* II, 16 is a good example of Aristotle relating a number of different projects from different parts of the *Organon*. The *Topics* and *S.E.* present his analysis of argumentation within the dialectical context of question and answer. *Pr. An.* I presents his formal theory of deduction through the syllogistic figures. The *Posterior Analytics* presents the special nature of deductions for the setting out (and teaching, see Barnes 1975a) of a scientific body of knowledge. Included in this latter sort of demonstration is epistemic justification (see Burnyeat 1981, esp. 126–33). In *Pr. An.* II, 16, Aristotle takes an error whose origin is in dialectic (Begging the Question), first shows how it violates the epistemic principles of scientific demonstrations, and then exemplifies the error through the various syllogistic figures.

3. *Pst. An.* I, 2, 71b19–22.

4. *Phy.* I, 1, 184a10–21; *Meta.* Z, 3, 1029b3–12; *EN* 1, 4, 1095a30–b4.

5. *Top.* I, 2, 101a34–b4. See Evans (1977), 31–41, on the role of dialectical question and answer in the discovery of scientific ἄρχαι.

6. Propositions here are not to be confused with the sentences that signify them. For Aristotle, qualities such as being better known, being prior, and being known through themselves belong not to language but to the things, events, or states of affairs signified by language.

7. Whether an Aristotelian first principle is one for which justification is merely unnecessary or actually impossible is a matter of dispute. I take it that Aristotle wants to make the stronger claim, that true primitives (e.g., the Law of Non-Contradiction) are impossible to justify by appeal to any other claims. At best one can only dialectically refute their denials. On the other hand, the ability to refute any and all denials of such a logical law would seem to constitute as strong of a justification for that logical law as one could desire. See *Meta.* Γ, 3–4.

8. *Pr. An.* II, 16, 64b36–38: ὅταν μὴ τὸ δι' αὑτοῦ γνωστὸν δι' αὑτοῦ τις ἐπιχειρῇ δείκνυναι, τότ' αἰτεῖται τὸ ἐξ ἀρχῆς.

9. Strictly speaking, this would not be "reasoning" (συλλογισμός). It amounts to a dialectical reminder to the audience of the self-explanatory nature of the claim. See p. 104 on Immediate Inferences.

10. For this reason I prefer to render Aristotle's distinction between things known δι' αὑτῶν and things known δι' ἄλλων by "self-explanatory" and "non-self-explana-

tory" rather than the more contemporary "self-evident" and "non-self-evident." For Aristotle, what is self-explanatory may not be self-evident to most people, and what appears self-evident to most people may be naturally non-self-explanatory.

11. Each is a necessary but not a sufficient condition for demonstration.

12. Aristotle suggests (ἴσως) without argumentation that what are γνωριμώτερα in the order of nature also are πιστώτερα to the properly trained and disposed mind. See *Top.* VI, 4, 142a2–11.

13. *Pr. An.* II, 16, 65a35–37; repeated at *Top.* VIII, 13, 162b31–33.

14. See, for example, Robinson (1971), who claims that the account of Begging the Question in the *Topics* is useless, because it consists in breaking a rule in "the Academic game of elenchus," which no one plays anymore. For the range of positions on the relationship between dialectic and scientific first principles in Aristotle, see Sim (1999).

15. *Top.* VIII, 5, 159a30–37: τῶν δ' ἀγωνιζομένων τὸν μὲν ἐρωτῶντα φαίνεσθαί τι δεῖ ποιεῖν πάντως, τὸν δ' ἀποκρινόμενον μηδὲν φαίνεσθαι πάσχειν· ἐν δὲ ταῖς διαλεκτικαῖς συνόδοις τοῖς μὴ ἀγῶνος χάριν ἀλλὰ πείρας καὶ σκέψεως τοὺς λόγους ποιουμένοις οὐ διήρθρωταί πω τίνος δεῖ στοχάζεσθαι τὸν ἀποκρινόμενον καὶ ποῖα διδόναι καὶ ποῖα μή, πρὸς τὸ καλῶς ἢ μὴ καλῶς φυλάττειν τὴν θέσιν)· ἐπεὶ οὖν οὐδὲν ἔχομεν παραδεδομένον ὑπ' ἄλλων, αὐτοί τι πειραθῶμεν εἰπεῖν.

16. *Top.* VIII, 13, 162b34–163a1: φανερώτατα μὲν καὶ πρῶτον, εἴ τις αὐτὸ τὸ δείκνυσθαι δέον αἰτήσειεν. τοῦτο δ' ἐπ' αὐτοῦ μὲν οὐ ῥᾳδιον λανθάνειν, ἐν δὲ τοῖς συνωνύμοις καὶ ἐν ὅσοις τὸ ὄνομα καὶ ὁ λόγος τὸ αὐτὸ σημαίνει, μᾶλλον.

17. See chapter 5, note 26.

18. Even if the distinction between Phenomenon 1 and Phenomenon 2 is untenable, Aristotle is still able to cite the false belief that some proposition p, under whatever description, is naturally self-explanatory as the ontological confusion that renders the false refutation persuasive.

19. *Top.* VIII, 13, 163a1–5.

20. *Top.* I, 12, 105a13–16: ἐπαγωγὴ δὲ ἡ ἀπὸ τῶν καθ' ἕκαστα ἐπὶ τὸ καθόλου ἔφοδος· οἷον εἰ ἔστι κυβερνήτης ὁ ἐπιστάμενος κράτιστος, καὶ ἡνίοχος, καὶ ὅλως ἐστὶν ὁ ἐπιστάμενος περὶ ἕκαστον ἄριστος.

21. *Top.* I, 12, 105a16–19; VIII, 2, 157a18–21.

22. The Latin commentators seem not to have been bothered by Aristotle's rendering of this third way of Begging the Question. They simply read the passage as disallowing arguments for universals by exhaustive enumeration of particulars, regardless of Aristotle's repeated praise elsewhere for such arguments. See William of Sherwood (1966), 157; Peter of Spain (1990), 147. My interpretation is supported by Basu (1986), 24–25.

23. *Top.* VIII, 16, 163a8–10.

24. *Top.* VIII, 16, 163a10–12: εἴ τις τῶν ἑπομένων ἀλλήλοις ἐξ ἀνάγκης θάτερον αἰτήσειεν, οἷον τὴν πλευρὰν ἀσύμμετρον εἶναι τῇ διαμέτρῳ, δέον ἀποδεῖξαι ὅτι ἡ διάμετρος τῇ πλευρᾷ.

25. τὰ ἑπόμενα ἀλλήλοις ἐξ ἀνάγκης is one of Aristotle's phrases for mutual entailment (i.e., logical equivalence).

26. Obversion is the denial of the contradictory of the original sentence. For example, "Some S's are P's" is logically equivalent to "It is not the case that all S's are P's," that is, "Some S's are non-P's."

27. *Pst. An.* I, 3, 73a10–11. Nor could one simply add an irrelevant premise to an immediate inference to produce a συλλογισμός. For example, the alleged argument: "Some S's are P's. Some S's are not P's. Therefore, some P's are S's" is, for Aristotle, only apparent reasoning, because it commits the fallacy of Non-Cause As Cause. See pp. 107ff. (The Fallacy of Treating a Non-Cause As a Cause).

28. Must the knowledge be conscious? Or can one not be aware that he knows the converse in knowing the original? If "being aware of" means "presently thinking of," then certainly one is not aware of most of what he knows at any given moment. But if a believer of "Some S's are P's" were to be asked whether he also believed "Some P's are S's" it would, for Aristotle, be psychologically impossible to sincerely answer "no." This poses no difficulty for Aristotle, because there can be no real doubters of the relationship. Conversion equivalences stand to human reason much as the Law of Non-Contradiction does: reasoning itself presupposes prior acceptance of the principles.

29. Ross (1949), 293; Smith (1989), 111–12.

30. Lear (1980), 3–6.

31. The contrapositive of a predication consists in predicating the negated subject term of the negated predicate term, for example, "All S's are P's" = "No non-P's are S's"; "Some S's are not P's" = "Some P's are not S's."

32. See e.g., *Top.* II, 8, 113b15–26. Patzig (1968) claims that "Aristotle never uses and never mentions" the laws of obversion and contraposition in his *Pr. An.* treatment of reduction (144). This is not because he failed to recognize their importance as self-explanatory equivalences. Rather, their use of negative terms made them nonprimitive. If they were primitive in the way that conversion equivalences are, there would be no need to resort to *reductio* proofs of the syllogistic forms of Baroco and Bocardo. The validity of Baroco can be shown directly by appeal to obversion and Ferio; the validity of Bocardo by appeal to contraposition, obversion, and Ferio.

33. An excellent example of a pair of logically equivalent contrapositives, one of which has a long history of philosophical acceptance, while the other has been almost universally rejected, is discussed by Burnyeat (1979). There he notes that it seems reasonable to many that

(a) All things that appear F to some observers and not-F to others do not possess F inherently.

But virtually no one supposes that

(b) All things that possess F inherently must appear F to all observers (or to no observers). Although (a) appears "more knowable to some," a refutation of (b) suffices to refute (a).

34. *Top.* VI, 4, 142a2–11.

35. *S.E.* 5, 167a36–39; 7, 169b12–17.

36. *S.E.* 6, 168b23–25: δεῖ γὰρ τὸ συμπέρασμα "τῷ ταῦτ' εἶναι" συμβαίνειν, ὅπερ οὐκ ἦν ἐν τοῖς ἀναιτίοις.

37. This is explained at *Pr. An.* II, 17, 65a38–66a15. Aristotle's back reference at 65b15–16 to his treatment of the fallacy "in the Topics" is a reference to our passages in *S.E.* Importantly, Aristotle nowhere in *S.E.* restricts Non-Cause fallacies to *reductio* arguments. They do, however, provide good examples of false refutations, because they lead to a denial of a previously accepted claim.

38. In fact, Aristotle's example below only leads to a falsehood, not a logical impossibility. The terminology of arguing εἰς τὸ ἀδύνατον here seems to be carried over from the *Prior Analytics,* where the technique is used to complete certain non-first figure syllogisms that would otherwise not be reducible to perfect first figure forms (*Pr. An.* I, 7, 29b1–6). In those examples (I, 5, 27a36–b3 [Baroco] and 28b17–20 [Bocardo]), the contradictory of the intended conclusion is assumed and shown to lead to the contradictory of an original premise. This contradictory, conjoined to the original premise, produces the logical impossibility. For a discussion of the role of this technique in Aristotle's syllogistic, see Smith (1989), 115–16.

39. See Moravcsik (1975).

40. *Top.* VI, 6, 145b16–20: ὁμοίως δὲ καὶ τῆς ἀπορίας δόξειεν ἂν ποιητικὸν εἶναι ἡ τῶν ἐναντίων ἰσότης λογισμῶν· ὅταν γὰρ ἐπ' ἀμφότερα λογιζομένοις ἡμῖν ὁμοίως ἅπαντα φαίνηται καθ' ἑκάτερον γίνεσθαι, ἀποροῦμεν ὁπότερον πράξωμεν. Also see *Top.* I, 13, 104b12–14.

41. Whether Aristotle intends (7) to be a mere restatement of (6) or a further inference from (6) is unclear. The issue is not important for the argument, for (6) no less than (7) is an obvious falsehood.

42. *S.E.* 5, 167b27–31: οἷον ὅτι οὐκ ἔστι ψυχὴ καὶ ζωὴ ταὐτόν, εἰ γὰρ φθορᾷ γένεσις ἐναντίον, καὶ τῇ τινὶ φθορᾷ ἔσται τὶς γένεσις· ὁ δὲ θάνατος φθορά τις καὶ ἐναντίον ζωῇ, ὥστε γένεσις ἡ ζωὴ καὶ τὸ ζῆν γίνεσθαι· τοῦτο δ' ἀδύνατον· οὐκ ἄρα ταὐτὸν ἡ ψυχὴ καὶ ἡ ζωή. Compare Plato's argument for the immortality of the soul at *Phaedo* 102–106.

43. For example, each particular kind of coming to be is contrary to one and only one particular kind of destruction, and so on.

44. Aristotle says that the argument (1) through (7) is not absolutely ἀσυλλόγιστος, but that it is ἀσυλλόγιστος to conclude the denial of (1).

45. *Top.* I, 1, 100a25–27: Ἔστι δὴ συλλογισμὸς λόγος ἐν ᾧ τεθέντων τινῶν ἕτερόν τι τῶν κειμένων ἐξ ἀνάγκης συμβαίνει διὰ τῶν κειμένων.

46. Moravcsik (1975), 622–38.

47. Hamblin (1970), 78.

48. *S.E.* 29, 181a31–35: Ὅσοι τε παρὰ τὸ προστιθέναι τι συλλογίζονται, σκοπεῖν εἰ ἀφαιρουμένου συμβαίνει μηδὲν ἧττον τὸ ἀδύνατον. κἄπειτα τοῦτο ἐμφανιστέον, καὶ λεκτέον ὡς ἔδωκεν οὐχ ὡς δοκοῦν ἀλλ' ὡς πρὸς τὸν λόγον,

ὁ δὲ κέχρηται οὐδὲν πρὸς τὸν λόγον. Aristotle's identification of the fallacy of Non-Cause here as "something additional being premissed" further confirms that relevancy is the issue at stake.

49. The wording of this line is similar to that found in *Top.* VIII, 6. There Aristotle notes that premises asked for by a dialectical questioner will possess two traits: (1) They will either seem true (i.e., be probable: ἔνδοξον), or seem false (i.e., be improbable: ἄδοξον), or seem neither; (2) They will either seem to be relevant to the argument or irrelevant (πρὸς τὸν λόγον ἢ μὴ πρὸς τὸν λόγον). Aristotle then surveys the six possible combinations of those features and offers explicit advice on how to answer a request for each type of premise. But none of his comments in *Top.* VIII, 6, seems appropriate to the issue of irrelevant premises in arguments "to the impossible." Such premises are rarely improbable. That, however, seems to be what is excused in *S.E.* 29 for the sake of their seeming relevance.

50. *Top.* VIII, 5, 159b17–20: θετέον οὖν τά τε δοκοῦντα πάντα καὶ τῶν μὴ δοκούντων ὅσα ἧττόν ἐστιν ἄδοξα τοῦ συμπεράσματος· ἱκανῶς γὰρ ἂν δόξειε διειλέχθαι.

51. *S.E.* 5, 167b29; 33: ὁ θάνατος ἐναντίον ζωῇ.

52. *Cat.* 10, 11b17ff. Cf. *Meta.* Δ, 10, 1018a20ff. Appeals to A and B as opposites (ἀντικείμενα) to show that they are contraries (ἐναντία) are examples of Begging the Question at *Top.* VIII, 13, 163a1–5, because contraries are a subset of opposites.

53. *Phy.* V, 1, 225a32–34.

54. For the distinguishing marks, see *Cat.* 10.

55. That is, not ἀσυλλόγιστος ἁπλῶς, says Aristotle at *S.E.* 5, 167b34.

56. A further reason for accepting the premise is the authority of Plato, who uses the same premise of life and death as ἐναντία at *Phaedo* 105d.

CHAPTER 7

1. *S.E.* 6, 168b27; 7, 169b6–7; 8, 170a4–5.

2. *S.E.* 5, 166b28–32: Οἱ μὲν οὖν παρὰ τὸ συμβεβηκὸς παραλογισμοί εἰσιν ὅταν ὁμοίως ὁτιοῦν ἀξιωθῇ τῷ πράγματι καὶ τῷ συμβεβηκότι ὑπάρχειν. ἐπεὶ γὰρ τῷ αὐτῷ πολλὰ συμβέβηκεν, οὐκ ἀνάγκη πᾶσι τοῖς κατηγορουμένοις καὶ καθ' οὗ κατηγορεῖται ταὐτὰ πάντα ὑπάρχειν.

3. *S.E.* 6, 168a40–b3: οὐδ' εἰ τὸ τρίγωνον δυοῖν ὀρθαῖν ἴσας ἔχει, συμβέβηκε δ' αὐτῷ σχήματι εἶναι ἢ πρώτῳ ἢ ἀρχῇ, ὅτι σχῆμα ἢ ἀρχὴ ἢ πρῶτον τοῦτό ἐστιν· οὐ γὰρ ᾗ σχῆμα οὐδ' ᾗ πρῶτον ἀλλ' ᾗ τρίγωνον ἡ ἀπόδειξις.

4. Aristotle uses "accident" (συμβεβηκός) or "accidental" in a variety of philosophical contexts. Its core feature is that of being non-necessary (*Meta.* Δ, 30, 1025a14–30). There is, however, a second use of "accidental" covered in *Metaphysics* Δ. Accidents also may be necessary features that are not part of the essence of their subjects.

Aristotle's example of this wider sense of "accident" is the possession by a triangle of angles equal to two right angles (*Meta.* Δ, 30, 1025a30–32). This would suggest that in the *S.E.* 6 argument the accidental premise is the first premise. Although this is true, Aristotle's subsequent discussion (see immediately following) makes clear that he regards the second premise as the cause of the false appearance of reasoning.

5. *S.E.* 24, 129a26–31: Πρὸς δὲ τοὺς παρὰ τὸ συμβεβηκὸς μία μὲν ἡ αὐτὴ λύσις πρὸς ἅπαντας. ἐπεὶ γὰρ ἀδιόριστόν ἐστι τὸ πότε λεκτέον ἐπὶ τοῦ πράγματος ὅταν ἐπὶ τοῦ συμβεβηκότος ὑπάρχῃ, καὶ ἐπ' ἐνίων μὲν δοκεῖ καὶ φασίν, ἐπ' ἐνίων δ' οὔ φασιν ἀναγκαῖον εἶναι, ῥητέον οὖν συμβιβασθέντος ὁμοίως πρὸς ἅπαντας ὅτι οὐκ ἀναγκαῖον, ἔχειν δὲ δεῖ προφέρειν τὸ "οἷον."

6. Bueno (1988), 12, has unjustly criticized Aristotle on this matter. She claims that Aristotle's "fallacy of accident is not a paralogism, since valid arguments can be constructed on its pattern." But the claim of the sophistical reasoner is that the conclusion necessarily follows. All that the answerer must show is that the sophist's conclusion does not necessarily follow. A single counterexample is sufficient to show this. Bueno seems to interpret "paralogism" as a necessarily false pattern of inference. But, for Aristotle, a paralogism is rather a pattern of inference that appears sound but is not. That appearance of soundness is generated by the fact that sometimes it is true that attributes belong in the same way to both subjects and their accidents.

7. Bueno (1988), 12. The same complaint can be found in Joseph (1916), 587–88.

8. *S.E.* 24, 179a37–38: φανερὸν γὰρ ἐν ἅπασι τούτοις ὅτι οὐκ ἀνάγκη τὸ κατὰ τοῦ συμβεβηκότος καὶ κατὰ τοῦ πράγματος ἀληθεύεσθαι· μόνοις γὰρ τοῖς κατὰ τὴν οὐσίαν ἀδιαφόροις καὶ ἓν οὖσιν ἅπαντα δοκεῖ ταὐτὰ ὑπάρχειν.

9. These are what McKirahan (1992) labels "*per se* 1" and "*per se* 2" predicates.

10. At *Pst. An.* I, 4, 73b4, Aristotle calls "accidents" those things that belong in neither of the two καθ' αὑτό senses. That is a weaker sense of "accidental" than the one I am reading here in *S.E.*, where what does not belong in both of the two senses is accidental.

11. Chapters 1, 2, and 3.

12. Bueno (1988), 11.

13. A third possibility, (∃x)(Mx & c ≠ x), is not relevant to the argument.

14. See e.g., Ross (1924), lxxxviii–lxxxix. Ross, however, is fully aware of the ontological basis of the confusion (306–308).

15. *de Int.* 3, 16b19–25.

16. See Sommers (1969) for the argument that Aristotle has no need for any special relation of identity to deal with individual terms. In the context of fallacies, Aristotle seems to take double καθ' αὑτό predication to be doing the necessary ontological work reflected by the identity relation.

17. *S.E.* 5, 166b35–36: διὰ τὸ συμβεβηκέναι οὗ ἔφησεν ἕτερον εἶναι, τοῦτον εἶναι ἄνθρωπον.

18. But as the modern analysis showed above, the appearance of validity in the third argument seems to require more than just the one mistaken accidental predication in the second premise.

19. *S.E.* 24, 179a32–37: εἰσὶ δὲ πάντες οἱ τοιοίδε τῶν λόγων παρὰ τὸ συμβεβηκός· "ἆρ' οἶδας ὃ μέλλω σε ἐρωτᾶν;" "ἆρ' οἶδας τὸν προσιόντα, ἢ τὸν ἐγκεκαλυμμένον;" . . . φανερὸν γὰρ ἐν ἅπασι τούτοις ὅτι οὐκ ἀνάγκη τὸ κατὰ τοῦ συμβεβηκότος καὶ κατὰ τοῦ πράγματος ἀληθεύεσθαι.

20. We already have seen Aristotle deal with some fallacies involving oblique contexts in his treatment of amphiboly (chapter 2). There, however, the fallacies arose because of syntactical and morphological features of Greek, for example, subjects in oblique contexts as well as objects taking the accusative case, or nominative and accusative neuter pronouns having identical morphology. Ambiguities such as τὸ βούλεσθαι λαβεῖν με τοὺς πολεμίους and ἆρ' ὅ τις γινώσκει, τοῦτο γινώσκει; do not involve purported identity predications.

21. *S.E.* 24, 179a39–b2: τῷ δ' ἀγαθῷ οὐ ταὐτόν ἐστιν ἀγαθῷ τ' εἶναι καὶ μέλλοντι ἐρωτᾶσθαι, οὐδὲ τῷ προσιόντι ἢ ἐγκεκαλυμμένῳ προσιόντι τε εἶναι καὶ Κορίσκῳ.

22. *S.E.* 20, 177b31–34.

23. *S.E.* 24, 179a35: ἆρα τὰ ὀλιγάκις ὀλίγα ὀλίγα;

24. Poste (1866), 156–57, offers the following analogous argument to illustrate his version of the sophism in Aristotle.

> Oxygen combined with hydrogen is water.
> Oxygen combined with hydrogen is oxygen.
> ___
> Therefore, oxygen is water.

Although Poste calls this an equivocation, it seems to be closer to a fallacy of Composition or Division. When one joins "oxygen-combined-with-hydrogen," the first statement is an identity. But the second premise must be divided differently to make sense: "Oxygen, [even when] combined with hydrogen, is oxygen." The difference between the two subject terms of the premises would be easily missed in written transcription without word breaks and sufficient punctuation. In speech, however, pauses in locution would reduce the likelihood of sophistic confusion.

25. Joseph (1916), 587.

26. For example, the unit of one house would belong καθ' αὑτό to the set of six houses as a line belongs to a triangle (*Posterior Analytics* sense I), but not as odd belongs to number (*Posterior Analytics* sense II).

27. *Cat.* 6, 5b14–22.

28. *S.E.* 24, 179b34–37: εἰ γάρ, μὴ συμπεραινομένου, τοῦτο παραλιπόντες ἀληθὲς συμπεπεράνθαι φασί (πάντα γὰρ εἶναι καὶ πολὺν καὶ ὀλίγον), ἁμαρτάνουσιν.

29. *S.E.* 24, 179b7–11: Λύουσι δέ τινες ἀναιροῦντες τὴν ἐρώτησιν. φασὶ γὰρ ἐνδέχεσθαι ταὐτὸ πρᾶγμα εἰδέναι καὶ ἀγνοεῖν, ἀλλὰ μὴ κατὰ ταὐτό· τὸν οὖν προσιόντα οὐκ εἰδότες, τὸν δὲ Κορίσκον εἰδότες, ταὐτὸ μὲν εἰδέναι καὶ ἀγνοεῖν φασιν, ἀλλ' οὐ κατὰ ταὐτό.

30. *Secundum Quid* is classified as a fallacy outside of language in Aristotle's scheme. In chapter 8 I argue that this position is not consistent with the rest of Aristotle's account of fallacies. These should be regarded as linguistically based fallacies whose resolution requires both linguistic and nonlinguistic clarifications.

31. *S.E.* 24, 179b11–16.

32. Aristotle makes reference to these two arguments without reproducing them. My reconstruction is based on the paternal dog sophism as it appears in Plato's *Euthydemus* 298e:

πατήρ ἐστιν ὁ κύων.
σός ἐστιν ὁ κύων.
____
ὥστε σὸς πατὴρ γίγνεται ὁ κύων.

33. *S.E.* 24, 179b26–33: ἴσως δὲ καὶ τοῦτ' ἐπ' ἐνίων οὐδὲν κωλύει συμβαίνειν· πλὴν ἐπί γε τούτων οὐδὲ τοῦτο δόξειεν ἄν· καὶ γὰρ τὸν Κορίσκον ὅτι Κορίσκος οἶδε καὶ τὸ προσιὸν ὅτι προσιόν. ἐνδέχεσθαι δὲ δοκεῖ τὸ αὐτὸ εἰδέναι καὶ μή, οἷον ὅτι μὲν λευκὸν εἰδέναι, ὅτι δὲ μουσικὸν μὴ γνωρίζειν· οὕτω γὰρ τὸ αὐτὸ οἶδε καὶ οὐκ οἶδεν, ἀλλ' οὐ κατὰ ταὐτόν. τὸ δὲ προσιὸν καὶ Κορίσκον <ὄν>, καὶ ὅτι προσιὸν καὶ ὅτι Κορίσκος, οἶδεν. This last line has been emended by Ross to read "But as to the one approaching *being* Coriscus . . ." Although this renders the line more readable, I fail to see how it makes the thought any more comprehensible.

34. Kirwan (1979), 45. See the next chapter for my discussion of *Secundum Quid.*

35.

That statue is yours.	That dog is yours.
That statue is a work.	That dog is a father.
______________	______________
That work is yours.	That father is yours.

36. *S.E.* 24, 179b38–180a7: Ἔνιοι δὲ καὶ τῷ διττῷ λύουσι τοὺς συλλογισμούς, οἷον ὅτι σός ἐστι πατὴρ ἢ υἱὸς ἢ δοῦλος. καίτοι φανερὸν ὡς εἰ παρὰ τὸ πολλαχῶς λέγεσθαι φαίνεται ὁ ἔλεγχος, δεῖ τοὔνομα ἢ τὸν λόγον κυρίως εἶναι πλειόνων. τὸ δὲ τόνδ' εἶναι τοῦδε τέκνον οὐδεὶς λέγει κυρίως, εἰ δεσπότης ἐστὶ τέκνου, ἀλλὰ παρὰ τὸ συμβεβηκὸς ἡ σύνθεσίς ἐστιν· "ἆρ' ἐστὶ τοῦτο σόν;" "ναί." "ἔστι δὲ τοῦτο τέκνον; σὸν ἄρα τοῦτο τέκνον, ὅτι συμβέβηκεν εἶναι καὶ σὸν καὶ τέκνον·" ἀλλ' οὐ σὸν τέκνον. With the Oxford and Loeb editors, I retain the mss. order of the last line, contra Wallies and Ross. I have adjusted the punctuation accordingly.

37. See Appendix 3 for a detailed discussion of this text.

38. But see chapter 8 below on *Secundum Quid* for Aristotle's reflections on the claim that the genitive case is linguistically ambiguous between its significations as possessive and as partitive.

39. *Cat.* 3, 1b10–15: Ὅταν ἕτερον καθ' ἑτέρου κατηγορῆται ὡς καθ' ὑποκειμένου, ὅσα κατὰ τοῦ κατηγορουμένου λέγεται, πάντα καὶ κατὰ τοῦ ὑποκειμένου ῥηθήσεται· οἷον ἄνθρωπος κατὰ τοῦ τινὸς ἀνθρώπου κατηγορεῖται, τὸ δὲ ζῷον κατὰ τοῦ ἀνθρώπου· οὐκοῦν καὶ κατὰ τοῦ τινὸς ἀνθρώπου τὸ ζῷον κατηγορηθήσεται· ὁ γὰρ τὶς ἄνθρωπος καὶ ἄνθρωπός ἐστι καὶ ζῷον.

40. See Gelber (1984, 1987); Boehner (1958), 201ff. The dispute continues among modern philosophers; see e.g., Trentman (1970).

41. Trentman (1970), 588. Trentman is particularly good at comparing William of Ockham's ideal of mental language with the ideal languages sought after by early twentieth-century linguistic philosophers.

42. For example, Hamblin (1970), 81, writes: "What does distinguish the refutations dependent on language is that they all arise from the fact that language is an imperfect instrument for the expression of our thoughts: the others could, in theory, arise even in a perfect language."

43. *S.E.* 5, 167b1–3: Ὁ δὲ παρὰ τὸ ἑπόμενον ἔλεγχος διὰ τὸ οἴεσθαι ἀντιστρέφειν τὴν ἀκολούθησιν· ὅταν γὰρ τοῦδε ὄντος ἐξ ἀνάγκης τόδε ᾖ, καὶ τοῦδε ὄντος οἴονται καὶ θάτερον εἶναι ἐξ ἀνάγκης.

44. See Hintikka (1973), 43–47.

45. For a selection of Aristotelian uses, see Ross (1949), 293.

46. For references, see Bonitz (1955), 66.

47. *S.E.* 5, 167b2–3.

48. *S.E.* 5, 167b5–6.

49. The practice led to several proverbial expressions (Liddell and Scott 1997a). See, e.g., the fragment of Diphilus in Edmonds (1961), vol. 3A, 136–37:

οὐκ ἀλλ' ἀλείψας τὴν τράπεζαν τῇ χολῇ
ὥσπερ τὰ παιδί' αὑτόν ἀπογαλακτιεῖ.

"No, but he applied some gall to the table, weaning himself like children from their mothers' milk."

50. *S.E.* 5, 167b4–5: αἱ περὶ τὴν δόξαν ἐκ τῆς αἰσθήσεως ἀπάται γίνονται.

51. *S.E.* 5, 167b6–8: καὶ ἐπεὶ συμβαίνει τὴν γῆν ὕσαντος γίνεσθαι διάβροχον, κἂν ᾖ διάβροχος, ὑπολαμβάνομεν ὗσαι. τὸ δ' οὐκ ἀναγκαῖον.

52. *Rhet.* I, 2, esp. 1357a22–b25; *Pr. An.* II, 27. In the *Rhetoric*, arguments from probable signs are assimilated to inductive reasoning, and arguments from necessary signs to deductive (I, 2, 1356a35–b11). In *S.E.*, however, these exemplify false συλλογισμοί, not false inductive arguments.

53. *S.E.* 5, 167b9–12.

54. The example with Socrates comes from Ps.-Alexander, 62, 25–29.

55. At *Phy.* I, 3, 186a11–16, Aristotle criticizes the argument of Melissus for committing two errors. The first is the fallacy due to Consequent. The second is the

assumption that just because something has a beginning in time, it also must have a beginning in space.

56. *S.E.* 5, 167b13–20: οἷον ὁ Μελίσσου λόγος ὅτι ἄπειρον τὸ ἅπαν, λαβὼν τὸ μὲν ἅπαν ἀγένητον (ἐκ γὰρ μὴ ὄντος οὐδὲν ἂν γενέσθαι), τὸ δὲ γενόμενον ἐξ ἀρχῆς γενέσθαι· εἰ μὴ οὖν γέγονεν, ἀρχὴν οὐκ ἔχειν τὸ πᾶν, ὥστ' ἄπειρον. οὐκ ἀνάγκη δὲ τοῦτο συμβαίνειν· οὐ γὰρ εἰ τὸ γενόμενον ἅπαν ἀρχὴν ἔχει, καὶ εἴ τι ἀρχὴν ἔχει, γέγονεν, ὥςπερ οὐδ' εἰ ὁ πυρέττων θερμός, καὶ τὸν θερμὸν ἀνάγκη πυρέττειν.

57. *S.E.* 28, 181a23–30: ἔστι δὲ διττὴ ἡ τῶν ἑπομένων ἀκολούθησις· ἢ γὰρ ὡς τῷ ἐν μέρει τὸ καθόλου, οἷον ἀνθρώπῳ ζῷον (ἀξιοῦται γάρ, εἰ τόδε μετὰ τοῦδε, καὶ τόδ' εἶναι μετὰ τοῦδε), ἢ κατὰ τὰς ἀντιθέσεις (εἰ γὰρ τόδε τῷδε ἀκολουθεῖ, τῷ ἀντικειμένῳ τὸ ἀντικείμενον)· παρ' ὃ καὶ ὁ τοῦ Μελίσσου λόγος· εἰ γὰρ τὸ γεγονὸς ἔχει ἀρχήν, τὸ ἀγένητον ἀξιοῖ μὴ ἔχειν, ὥστ' εἰ ἀγένητος ὁ οὐρανός, καὶ ἄπειρος. τὸ δ' οὐκ ἔστιν· ἀνάπαλιν γὰρ ἡ ἀκολούθησις.

58. *S.E.* 6, 168b27–40: Οἱ δὲ παρὰ τὸ ἑπόμενον μέρος εἰσὶ τοῦ συμβεβηκότος· τὸ γὰρ ἑπόμενον συμβέβηκε. διαφέρει δὲ τοῦ συμβεβηκότος, ὅτι τὸ μὲν συμβεβηκὸς ἔστιν ἐφ' ἑνὸς μόνου λαβεῖν, οἷον ταὐτὸ εἶναι τὸ ξανθὸν καὶ μέλι, καὶ τὸ λευκὸν καὶ κύκνον, τὸ δὲ παρὰ τὸ ἑπόμενον ἀεὶ ἐν πλείοσιν· τὰ γὰρ ἑνὶ καὶ ταὐτῷ ταὐτὰ καὶ ἀλλήλοις ἀξιοῦμεν εἶναι ταὐτά· διὸ γίνεται παρὰ τὸ ἑπόμενον ἔλεγχος. ἔστι δ' οὐ πάντως ἀληθές, οἷον ἂν ᾖ κατὰ συμβεβηκός· καὶ γὰρ ἡ χιὼν καὶ ὁ κύκνος τῷ λευκῷ ταὐτόν. ἢ πάλιν, ὡς ἐν τῷ Μελίσσου λόγῳ, τὸ αὐτὸ εἶναι λαμβάνει τὸ γεγονέναι καὶ ἀρχὴν ἔχειν . . . ὅτι γὰρ τὸ γεγονὸς ἔχει ἀρχήν, καὶ τὸ ἔχον ἀρχὴν γεγονέναι ἀξιοῖ, ὡς ἄμφω ταὐτὰ ὄντα τῷ ἀρχὴν ἔχειν, τό τε γεγονὸς καὶ τὸ πεπερασμένον.

59. Bueno (1988), 11.

60. Both medieval Greek commentaries emphasize this connection to convertibility. See Ps.-Alexander (1898), 62–64; Sophonias (1884), 19.

61. Aristotle does not actually formulate this argument from *S.E.* 6, 168a40–b3, with universals. If he only means to apply the argument to some particular triangle, then this example fails to be a problem for the theory that fallacies due to Accident involve a particular accidental premise and fallacies due to Consequent involve a universal accidental premise.

62. Actually, all that is needed for the fallacy is that one of the premises be accidental.

63. It is unclear which conclusions (3a or 3b) are supposed to be falsely derived from the accidental premise. In the Melissus argument, if the fallaciously derived conclusion is (3a), then its subsequent use with the premise "the universe is ungenerated" validly entails that the universe is infinite by proper conversion by opposites. If the fallaciously derived conclusion is (3b), then there remains the improper conversion by opposites to reach Melissus's desired conclusion. In the latter case, Aristotle ends up with two errors due to Consequent in Melissus's argument. In the former case, the error due to Consequent in *S.E.* 5 and 28 turns out to be a different error from this one in *S.E.* 6. Here the way one derives "what is finite is generated" is by a separate false argument; in the discussions of *S.E.* 5 and 28, the way one derives it is simply by falsely converting by opposites.

CHAPTER 8

1. *S.E.* 5, 166b37–167a1: Οἱ δὲ παρὰ τὸ ἁπλῶς τόδε ἢ πῇ λέγεσθαι καὶ μὴ κυρίως, ὅταν τὸ ἐν μέρει λεγόμενον ὡς ἁπλῶς εἰρημένον ληφθῇ.

2. See Kirwan (1979).

3. See chapter 1 for a discussion of this distinction.

4. *S.E.* 5, 167a1: εἰ τὸ μὴ ὄν ἐστι δοξαστόν, ὅτι τὸ μὴ ὂν ἔστιν.

5. It is another question whether Aristotle believes such predications can be true or false.

6. *S.E.* 5, 167a2–4: οὐ γὰρ ταὐτὸ τὸ εἶναί τέ τι καί εἶναι ἁπλῶς . . . οὐ γὰρ ταὐτὸ τὸ μὴ εἶναί τι καὶ ἁπλῶς μὴ εἶναι.

7. *S.E.* 5, 167a4–6: φαίνεται δὲ διὰ τὸ πάρεγγυς τῆς λέξεως καὶ μικρὸν διαφέρειν τὸ εἶναί τι τοῦ εἶναι, καὶ τὸ μὴ εἶναί τι τοῦ μὴ εἶναι. See also *S.E.* 7, 169b9–12.

8. *S.E.* 6, 168b11–16: Οἱ δὲ παρὰ τὸ πῇ καὶ ἁπλῶς, ὅτι οὐ τοῦ αὐτοῦ ἡ κατάφασις καὶ ἡ ἀπόφασις. τοῦ γὰρ πῇ λευκοῦ τὸ πῇ οὐ λευκόν, τοῦ δ' ἁπλῶς λευκοῦ τὸ ἁπλῶς οὐ λευκὸν ἀπόφασις· εἰ οὖν δόντος πῇ εἶναι λευκὸν ὡς ἁπλῶς εἰρημένου λαμβάνει, οὐ ποιεῖ ἔλεγχον, φαίνεται δὲ διὰ τὴν ἄγνοιαν τοῦ τί ἐστιν ἔλεγχος.

9. *S.E.* 25, 180a26–31: τὰ γὰρ ἐναντία καὶ τὰ ἀντικείμενα καὶ φάσιν καὶ ἀπόφασιν ἁπλῶς μὲν ἀδύνατον ὑπάρχειν τῷ αὐτῷ, πῇ μέντοι ἑκάτερον ἢ πρός τι ἢ πώς, ἢ τὸ μὲν πῇ τὸ δ' ἁπλῶς, οὐδὲν κωλύει. ὥστ' εἰ τόδε μὲν ἁπλῶς τόδε δὲ πῇ, οὔπω ἔλεγχος, τοῦτο δ' ἐν τῷ συμπεράσματι θεωρητέον πρὸς τὴν ἀντίφασιν.

10. See Kirwan (1979).

11. *S.E.* 5, 167a29–34.

12. *S.E.* 5, 167a35: ἕλκοι δ' ἄν τις τοῦτον καὶ εἰς τοὺς παρὰ τὴν λέξιν.

13. One might compare Plato's use of ὅλκη in *Philebus* 57d3–4, where the issue is again one of classification. Socrates argues for distinctions among kinds of knowledge, despite the single name possessed by all, and despite the arguments to the contrary by "those clever concerning the twisting (or forcing) of arguments" (τοῖς δεινοῖς περὶ λόγων ὁλκήν).

14. See Poste (1866), 110; Evans (1975), 51–52.

15. I note here a third possibility which, however, contains too many problems to qualify as a serious interpretation candidate. One might argue that the reference be restricted just to the immediately preceding subtype of *Secundum Quid* confusions, namely, those that fail to deny the same predicate in the same way at the same time. The warrant for thinking that Aristotle may be thinking here solely of fallacies hinging on confused temporal qualifiers is his earlier claim (see chapter 2) that general names that lack definite time references (e.g., "the sitting person") are homonymous and thereby can result in fallacious reasoning due to language. Perhaps Aristotle is

thinking that an argument that hinges upon a failure to make clear temporal qualifications is not unlike a homonymously indefinite reference, and so also might be considered due to language. If this were Aristotle's thinking, one would need to read ἕλκω as allowing the alternative analysis rather than as outright rejecting it. This, however, is unlikely to be Aristotle's meaning. Elsewhere he offers no hint of accounting for the generation of fallacies of *Secundum Quid* by linguistic means, and his examples in the *Topics* include instances of temporal qualifiers being omitted: see, e.g., *Top*. II, 11, 115b26–27.

16. *S.E.* 24, 180a8–10: Καὶ τὸ εἶναι τῶν κακῶν τι ἀγαθόν· "ἡ γὰρ φρόνησίς ἐστιν ἐπιστήμη τῶν κακῶν." τὸ δὲ τοῦτο τούτων εἶναι οὐ λέγεται πολλαχῶς, ἀλλὰ κτῆμα. Interestingly, according to Poste (157), subsequent grammarians in antiquity recognized the partitive use of the genitive as being its original meaning!

17. It is important in translation to catch the way that Aristotle introduces this objection at *S.E.* 180a10: εἰ δ' ἄρα πολλαχῶς. He intends this to be an improbable or an unbelieved possibility. See Smyth (1920), 2796.

18. *S.E.* 24, 180a11–13: καὶ γὰρ τὸν ἄνθρωπον τῶν ζῴων φαμὲν εἶναι, ἀλλ' οὔ τι κτῆμα· καὶ ἐάν τι πρὸς τὰ κακὰ λέγηται ὡς τινός, διὰ τοῦτο τῶν κακῶν ἐστιν, ἀλλ' οὐ τοῦτο τῶν κακῶν. My translation of the last line brings out the distinction that Aristotle intends. The two Greek phrases are identical, of course, which is the whole point of the objection.

19. *S.E.* 24, 180a13–14: παρὰ τὸ πῇ οὖν καὶ ἁπλῶς φαίνεται.

20. *S.E.* 24, 180a14–18: καίτοι ἐνδέχεται ἴσως ἀγαθὸν εἶναί τι τῶν κακῶν διττῶς, ἀλλ' οὐκ ἐπὶ τοῦ λόγου τούτου, ἀλλ' εἴ τι δοῦλον εἴη ἀγαθὸν μοχθηροῦ, μᾶλλον. ἴσως δ' οὐδ' οὕτως· οὐ γὰρ εἰ ἀγαθὸν καὶ τούτου, ἀγαθὸν τούτου ἅμα.

21. Aristotle prefaces his objection to this resolution at *S.E.* 180a16 with 'ἴσως. This does not, I believe, signal a lack of certainty about Aristotle's conclusion. He also prefaces the alternative resolution with 'ἴσως two lines earlier. He is ruminating: "On the one hand there appears this argument for there being double meaning here. But on the other hand there appears this objection to that argument." His belief that the objection wins the day is clear from the next line: "nor (οὐδέ) is 'man is of the animals' said in many ways."

22. See my treatment of this example in chapter 7.

23. *S.E.* 24, 180a18–22: οὐδὲ τὸ τὸν ἄνθρωπον φάναι τῶν ζῴων εἶναι [οὐ] λέγεται πολλαχῶς· οὐ γὰρ εἴ ποτέ τι σημαίνομεν ἀφελόντες, τοῦτο λέγεται πολλαχῶς· καὶ γὰρ τὸ ἥμισυ εἰπόντες τοῦ ἔπους "δός μοι Ἰλιάδα" σημαίνομεν, οἷον τὸ "μῆνιν ἄειδε, θεά."

24. In *Meta*. Z, 4, 1030a27–b3, Aristotle claims that to predicate being (ὄντα) of both substances and members of nonsubstance Categories is either to use the predicate homonymously or to use it "by adding to or taking away (προστιθέντας καὶ ἀφαιροῦντας) just as that which is unknown can also be said to be known [to be unknown]." Aristotle proceeds to deny that the predicate is either homonymous or the same, but like "medical," it is related to one and the same thing (πρὸς ἕν). It is unclear in this passage whether Aristotle is identifying πρὸς ἕν predication with

predication "by adding to or taking away." I merely note that "adding to or taking away" (i.e., *Secundum Quid*) is here distinguished from homonymy. On the other hand, we have seen that Aristotle works with both wider and narrower notions of homonymy in *S.E.* Kirwan's challenge remains: Aristotle has already used "homonymy" to cover cases naturally understood as involving predicates differently qualified.

25. *S.E.* 7, 169b9–12: καὶ τῶν παρὰ τὸ πῂ καὶ ἁπλῶς ἐν τῷ παρὰ μικρὸν ἡ ἀπάτη· ὡς γὰρ οὐδὲν προσσημαῖνον τὸ τὶ ἢ πῂ ἢ τὸ πὼς ἢ τὸ νῦν καθόλου συγχωροῦμεν.

26. *S.E.* 5, 167a2–6.

27. *S.E.* 25, 180a23–24: ἢ πῂ ἢ ποὺ ἢ πὼς ἢ πρός τι.

28. It might even be possible to show that all *Secundum Quid* fallacies involve Category mistakes.

29. *S.E.* 6, 169a18–21: ὥστε πάντες οἱ τόποι πίπτουσιν εἰς τὴν τοῦ ἐλέγχου ἄγνοιαν, οἱ μὲν οὖν παρὰ τὴν λέξιν, ὅτι φαινομένη <ἡ> ἀντίφασις, ὅπερ ἦν ἴδιον τοῦ ἐλέγχου, οἱ δ' ἄλλοι παρὰ τὸν τοῦ συλλογισμοῦ ὅρον.

30. *Top.* I, 100a25–27; *S.E.* 12, 165a1–2.

CHAPTER 9

1. *S.E.* 5, 167b38–168a1: οἱ δὲ παρὰ τὸ τὰ δύο ἐρωτήματα ἓν ποιεῖν, ὅταν λανθάνῃ πλείω ὄντα καὶ ὡς ἑνὸς ὄντος ἀποδοθῇ ἀπόκρισις μία.

2. *S.E.* 6, 169a15–16.

3. ἓν καθ' ἑνός. See e.g., *de Int.* 8, 18a13; *S.E.* 6, 169a7–8.

4. The Revised Oxford translation footnotes this reference at *de Int.* 11, 20b26, with "See esp. *Topics* VIII." Ackrill (1963), 145, also refers the reader to *Topics* VIII, as well as to our texts in *S.E.* I can find nothing in *Topics* VIII directly pertinent to the issues of the individuation of premises or the unity of terms. It seems to me that the reference can only be to the *S.E.* passages.

5. *S.E.* 30, 181a37–39: ἐρώτησις γὰρ μία πρὸς ἣν μία ἀπόκρισις ἔστιν, ὥστ' οὔτε πλείω καθ' ἑνὸς οὔτε ἓν κατὰ πολλῶν, ἀλλ' ἓν καθ' ἑνὸς φατέον ἢ ἀποφατέον. "For a single question is that to which there is a single answer, so that one must not affirm or deny several things of one, nor one of many, but one of one."

6. *S.E.* 5, 168a2–3: πότερον ἡ γῆ θάλαττά ἐστιν ἢ ὁ οὐρανός;

7. The fact that the question already assumes what is false is perhaps the only similarity that Aristotle's example has to the modern textbook treatment of the fallacy called "Many Questions." The classic modern example is the question "Have you stopped beating your wife?" There is considerable interesting literature on what exactly is wrong with such a question. See Walton (1981), Fair (1973), Hamblin (1970), 215–18, and their respective bibliographies. On an informal level, the problem with the question is that it presupposes an affirmative response to a prior question "Did you used to beat your wife?" Embedded in the "multiple question," then, is not just a

second question, but also a predetermined answer to that second question. It has, then, this similarity to Aristotle's disjunctive question: if one answers the question in either of its permissible ways, one commits oneself to a falsehood. In Aristotle's example, the falsehood is the answer itself. In the modern example, the falsehood is more deeply embedded in a presupposition to the answer.

8. *S.E.* 5, 168a1–2.

9. *S.E.* 30, 181a37–38: ἐρώτησις γὰρ μία πρὸς ἣν μία ἀπόκρισις ἔστιν. This view that the unity of a premise-question is determined by the possibility of a single answer is a local example of the same methodology that established the Principle of Parsimony: the unity of a fallacy-type is determined by the possibility of a single resolution.

10. *S.E.* 5, 168a7–11: ὧν τὰ μέν ἐστιν ἀγαθὰ τὰ δ' οὐκ ἀγαθά, πάντα ἀγαθὰ ἢ οὐκ ἀγαθά; ὁπότερον γὰρ ἂν φῇ, ἔστι μὲν ὡς ἔλεγχον ἢ ψεῦδος φαινόμενον δόξειεν ἂν ποιεῖν· τὸ γὰρ φάναι τῶν μὴ ἀγαθῶν τι εἶναι ἀγαθὸν ἢ τῶν ἀγαθῶν μὴ ἀγαθὸν ψεῦδος. For another version of the same argument, see *S.E.* 30, 181b9–13.

11. *S.E.* 5, 168a5–7: οἷον ἆρ' οὗτος καὶ οὗτός ἐστιν ἄνθρωπος; ὥστ' ἄν τις τύπτῃ τοῦτον καὶ τοῦτον, ἄνθρωπον ἀλλ' οὐκ ἀνθρώπους τυπτήσει.

12. *S.E.* 3. See my discussion (Form of the Expression and Solecism: Aristotle and Protagoras) in chapter 3, pp. 48ff.

13. So Ps.-Alexander has it (p. 53). He restates the example by replacing οὗτος καὶ οὗτος with ὁ Σωκράτης καὶ ὁ Πλάτων.

14. So Peter of Spain has it (p. 158). He illustrates the argument by having the first οὗτος refer to Socrates and the second οὗτος refer to Brunellus the ass, a favorite example in medieval pedagogy.

15. *S.E.* 5, 168a11–17: ὁτὲ δὲ προσληφθέντων τινῶν κἂν ἔλεγχος γίνοιτο ἀληθινός, οἷον εἴ τις δοίη ὁμοίως ἓν καί πολλὰ λέγεσθαι λευκὰ καὶ γυμνὰ καὶ τυφλά. εἰ γὰρ τυφλὸν τὸ μὴ ἔχον ὄψιν πεφυκὸς δ' ἔχειν, καὶ τυφλὰ ἔσται τὰ μὴ ἔχοντα ὄψιν πεφυκότα δ' ἔχειν. ὅταν οὖν τὸ μὲν ἔχῃ τὸ δὲ μὴ ἔχῃ, τὰ ἄμφω ἔσται ἢ ὁρῶντα ἢ τυφλά· ὅπερ ἀδύνατον.

16. Emphasis added. *S.E.* 30, 181b1–7: ὁτὲ μὲν ἀμφοῖν ὁτὲ δ' οὐδετέρῳ ὑπάρχει, ὥστε μὴ ἁπλοῦ ὄντος τοῦ ἐρωτήματος ἁπλῶς ἀποκρινομένοις οὐδὲν συμβαίνει πάσχειν . . . ὅταν μὲν οὖν τὰ πλείω τῷ ἑνὶ ἢ τὸ ἓν τοῖς πολλοῖς ὑπάρχῃ, τῷ ἁπλῶς δόντι καὶ ἁμαρτόντι ταύτην τὴν ἁμαρτίαν οὐδὲν ὑπεναντίωμα συμβαίνει, ὅταν δὲ τῷ μὲν τῷ δὲ μή, ἢ πλείω κατὰ πλειόνων.

17. "X is and Y is a man" is equivalent in signification to "X is a man, and Y is a man" (*de Int.* 8, 18a21–23), which is equivalent to "X and Y are men."

18. *S.E.* 30, 181a36–37: Πρὸς δὲ τοὺς τὰ πλείω ἐρωτήματα ἓν ποιοῦντας εὐθὺς ἐν ἀρχῇ διοριστέον.

19. *S.E.* 17, 176a9–12: οὐ γὰρ εἰ ἀληθὲς εἰπεῖν, διὰ τοῦτο μία ἡ ἐρώτησις. ἐγχωρεῖ γὰρ καὶ μυρία ἕτερα ἐρωτηθέντα ἐρωτήματα ἁπλῶς ἢ "ναί" ἢ "οὔ" ἀληθὲς εἶναι λέγειν· ἀλλ' ὅμως οὐκ ἀποκριτέον μιᾷ ἀποκρίσει· ἀναιρεῖται γὰρ

τὸ διαλέγεσθαι. Translations can mislead by their rendering of ἀναιρεῖται γὰρ τὸ διαλέγεσθαι at 176a12–13. The Loeb renders it "the ruin of discussion," which is too broad. Poste translates it as "there could be no dialectic," which is too narrow if dialectical reasoning is being contrasted to the other forms of reasoning. Better is the Revised Oxford rendering, "the death of argument," as long as "argument" is understood in the sense of συλλογίζεσθαι, which I consider the closest synonym to διαλέγεσθαι in this context.

20. *S.E.* 2, 165a38–39: Ἔστι δὴ τῶν <u>ἐν τῷ διαλέγεσθαι</u> λόγων τέτταρα γένη, διδασκαλικοὶ καὶ διαλεκτικοὶ καὶ πειραστικοὶ καὶ ἐριστικοί (emphasis added). See also *Topics* I, 1, and my discussion in chapter 1.

21. *S.E.* 17, 175b39–176a9: Εἰ δὲ τὰ δύο ἐρωτήματα μὴ ἓν ἐποίει τις ἐρώτημα, οὐδ' ἄν ὁ παρὰ τὴν ὁμωνυμίαν καὶ τὴν ἀμφιβολίαν ἐγίνετο παραλογισμός, ἀλλ' ἢ ἔλεγχος ἢ οὔ. τί γὰρ διαφέρει ἐρωτῆσαι εἰ Καλλίας καὶ Θεμιστοκλῆς μουσικοί εἰσιν ἢ εἰ ἀμφοτέροις ἓν ὄνομα ἦν ἑτέροις οὖσιν; εἰ γὰρ πλείω δηλοῖ ἑνός, πλείω ἠρώτησεν. εἰ οὖν μὴ ὀρθὸν πρὸς δύο ἐρωτήσεις μίαν ἀπόκρισιν ἀξιοῦν λαμβάνειν ἁπλῶς, φανερὸν ὅτι οὐδενὶ προσήκει τῶν ὁμωνύμων ἀποκρίνεσθαι ἁπλῶς, οὐδ' εἰ κατὰ πάντων ἀληθές, ὥσπερ ἀξιοῦσί τινες. οὐδὲν γὰρ τοῦτο διαφέρει ἤ εἰ ἤρετο, Κορίσκος καὶ Καλλίας πότερον οἴκοι εἰσὶν ἢ οὐκ οἴκοι, εἴτε παρόντων ἀμφοῖν εἴτε μὴ παρόντων· ἀμφοτέρως γὰρ πλείους αἱ προτάσεις.

22. *S.E.* 30, 181b9–13.

23. *S.E.* 30, 181b13–15: εἰ ἕκαστον αὐτὸ αὑτῷ ταὐτὸ καὶ ἄλλου ἕτερον, ἐπειδὴ οὐκ ἄλλοις ταὐτὰ ἀλλ' αὑτοῖς καὶ ἕτερα αὐτῶν, τὰ αὐτὰ ἑαυτοῖς ἕτερα καὶ τὰ αὐτά.

24. *S.E.* 30, 181b19–21: Ἐμπίπτουσι μὲν οὖν οὗτοι καὶ εἰς ἄλλας λύσεις· καὶ γὰρ τὸ ἄμφω καὶ τὸ ἅπαντα πλείω σημαίνει· οὔκουν ταὐτόν, πλὴν ὄνομα, συμβαίνει φῆσαι καὶ ἀποφῆσαι.

25. A careful Greek speaker is able to avoid these potential ambiguities with ἄμφω and ἅπαντα by using ἑκάτερον and ἕκαστον to refer to the separate members of a plurality.

26. Caution is needed here, for syncategorematic words such as "both" and "all" do not signify in the strict Aristotelian sense of σημαίνειν.

27. *Cael.* I, 11, 280b2–6: πολλαχῶς γὰρ λεγομένων, κἂν μηδὲν διαφέρῃ πρὸς τὸν λόγον, ἀνάγκη τὴν διάνοιαν ἀορίστως ἔχειν, ἄν τις τῷ διαιρουμένῳ πολλαχῶς ὡς ἀδιαιρέτῳ χρῆται· ἄδηλον γὰρ κατὰ ποίαν φύσιν αὐτῷ συμβαίνει τὸ λεχθέν.

28. *de Int.* 5, 17a15–17: ἔστι δὲ εἷς λόγος ἀποφαντικὸς ἢ ὁ ἓν δηλῶν ἢ ὁ συνδέσμῳ εἷς, πολλοὶ δὲ οἱ πολλὰ καὶ μὴ ἓν ἢ οἱ ἀσύνδετοι. For the translation of δηλοῦν by "signify," see Irwin (1982), 243–44 and note 6 to chapter 2.

29. Ackrill (1963), 125–27.

30. One possible defense of Aristotle would be to point to his distinction made several lines later in *de Int.* 5 between a simple (ἁπλῆ) statement and a compound

(συγκειμένη) statement, where the former expressly lacks any connectives (17a20–22). But this division seems to be drawn *among* the "single statement-making sentences," still suggesting a sense of unity to compound propositions not countenanced later in *S.E.*

31. *de Int.* 8, 18a25–26: οὐ γάρ ἐστιν τὶς ἄνθρωπος ἵππος.

32. *de Int.* 11, 20b12–26.

33. *de Int.* 5, 17a13–15.

34. See e.g., Z, 12; H, 6; I, 9, and elsewhere.

35. See e.g., Ackrill (1963), 145.

36. That error is more likely to result in fallacies due to Accident.

Conclusion and Summary

1. *S.E.* 1, 165a34–37; 34, 184a1–8.

Appendix 1

1. *S.E.* 1, 164a20-22: Περὶ δὲ τῶν σοφιστικῶν ἐλέγχων καὶ τῶν φαινομένων μὲν ἐλέγχων, ὄντων δὲ παραλογισμῶν ἀλλ' οὐκ ἐλέγχων, λέγωμεν ἀρξάμενοι κατὰ φύσιν ἀπὸ τῶν πρώτων.

2. Poste (1866), 97, 120-121.

3. See *S.E.* 1, 164a20-21, and perhaps *S.E.* 8, 170a9-11, which concludes that the sources of (1) are exactly the same as the sources of (2).

Appendix 2

1. *Republic* VI, 487b1-d5.

2. There may be less alteration in the meaning of "counters" than at first appears. Recent attempts to reconstruct the nature of this ancient "checkers" game have led to the theory that it may have been the precursor to the medieval game of Rithmomachia. In this game of the "Battle of Numbers," each counter represented a specific numeral, and victory was secured by maneuvering one's numerals into various desirable arithmetical, geometrical, and harmonic sequences. For the rules of Rithmomachia, see Smith and Eaton (1911). Specific references to the connection between arithmetic and a board game occur throughout Plato: e.g., *Statesman* 299e; *Gorgias* 450d; *Laws* 820c–d. See review article of recent literature on Rithmomachia by Artmann (1989).

3. It might appear to be an important difference that the experienced sophist in Aristotle's analogy intends to wreak confusion in the mind of the reasoner, whereas

the experienced Socrates in Plato's analogy does not intend confusion. But is this really the case? Socrates, too, intends the confusion in the minds of his interlocutors as propaedeutic to their obtaining true knowledge. The difference is that Socrates is portrayed as using proper reasoning to confute his interlocutor's false beliefs, while Aristotle's sophist is portrayed as using false reasoning to confute his fellow dialectician's true belief.

APPENDIX 3

1. *S.E.* 4, 166a14-23: εἰσὶ δὲ τρεῖς τρόποι τῶν παρὰ τὴν ὁμωνυμίαν καὶ τὴν ἀμφιβολίαν· εἷς μὲν ὅταν ἢ ὁ λόγος ἢ τοὔνομα κυρίως σημαίνῃ πλείω, οἷον ἀετὸς καὶ κύων· εἷς δὲ ὅταν εἰωθότες ὦμεν οὕτω λέγειν· τρίτος δὲ ὅταν τὸ συντεθὲν πλείω σημαίνῃ, κεχωρισμένον δὲ ἁπλῶς. οἷον τὸ "ἐπίσταται γράμματα"· ἑκάτερον μὲν γάρ, εἰ ἔτυχεν, ἕν τι σημαίνει, τὸ "ἐπίσταται" καὶ τὸ "γράμματα"· ἄμφω δὲ πλείω, ἢ τὸ τὰ γράμματα αὐτὰ ἐπιστήμην ἔχειν ἢ τῶν γραμμάτων ἄλλον. Ἡ μὲν οὖν ἀμφιβολία καὶ ὁμωνυμία παρὰ τούτους τοὺς τρόπους ἐστίν.

2. They do this by "imitating the nature itself of the thing named" (*Cratylus* 423a: μιμούμενοι αὐτὴν τὴν φύσιν τοῦ πράγματος.) For example, the name of Zeus "makes clear the nature of the god which, indeed, is what we said a name should be able to do" (*Cratylus* 396a). See also *Cratylus* 422d–e.

3. *Cratylus* 383a7-b2: ὀρθότητά τινα τῶν ὀνομάτων πεφυκέναι καὶ Ἕλλησι καὶ βαρβάροις τὴν αὐτὴν ἅπασιν.

4. *Cratylus* 433e.

5. For a brief review of the debate, see Robins (1951), ch. 1. A more extensive treatment occurs in W. S. Allen (1948).

6. See my note 15 in chapter 1. In addition to Kahn (1973), I have benefited from Kirk (1951), D. J. Allen (1954), Robinson (1969a, 1969b), and Ware (1987).

7. 389b–c.

8. *de Int.* 2, 16a26–28; 4, 16b33–17a2.

9. *Poet.* 21, 1457b1–6.

10. *HA* 5, 5, 540b18; 6, 11, 566a31. αἐτός also was a standard name for an architectural pediment. Perhaps the common use of eagles as decorative parts of pediments led to this eventual use by metonymy of αἐτός for the entire pediment. The origin of standard names is not important to Aristotle's theory.

11. *Rhet.* II, 24, 1401a15.

12. *Poet.* 21, 1457b25–30.

13. *Top.* I, 7, 103a23–31.

14. At this point it is important to acknowledge that I am equating two different descriptions Aristotle makes of this project of making distinctions κυρίως. In the

Poetics and *Rhetoric* passages, Aristotle describes the search for standard names of some given extralinguistic referent. Elsewhere, the search is described as one for the standard referent for some given name. I understand these to be two different approaches, dictated by differing contexts, to the same project. Aristotle divides areas of knowledge (ἐπιστήμη) or thought (διάνοια) into three broad types: those directed toward production (ποιητική), those directed toward action (πρακτική), and those directed toward speculation (θεωρητική). (See *Top.* VI, 6, 145a15–18, and *Meta.* E, 1, 1025b25.) The emphasis on the search for standard names is appropriate to the *Poetics* and *Rhetoric*, both being studies of certain productive arts of literary composition. The emphasis on the search for standard significates of names characterizes the speculative (and, to a lesser degree, the practical) sciences. Having noted this, it is interesting that κυρίως distinctions in the *Topics* and *S.E.* are typically put in the mode appropriate to the speculative rather than the productive arts. This has not been sufficiently appreciated by those who continue to regard Aristotelian dialectic as merely a skill in producing arguments.

15. The most developed case for the methodological importance of such an appeal—and its metaphysical underpinnings—is in Evans (1977), ch. 3. My brief account is heavily influenced by Evans's more detailed treatment.

16. See chapter 8 on *Secundum Quid.*

17. The further question of how one recognizes such experts is a separate, though still important, issue. Aristotle's generally optimistic view of the human person's ability to recognize the true and the good made this less problematic for him than it is for most modern thinkers.

18. *S.E.* 7, 169a31–33: ποῖα ὡσαύτως καὶ ποῖα ὡς ἑτέρως λέγεται (σχεδὸν γὰρ ὁ τοῦτο δυνάμενος ποιεῖν ἐγγύς ἐστι τοῦ θεωρεῖν τἀληθές . . .).

19. It should be noted that the bare phrase ἐπίσταται γράμματα is called amphibolous, independent of any context. This might seem to indicate an apparent disregard for context, which was found to be one distinctive criterion of amphiboly. On the other hand, the mere conjunction of the noun (having the same nominative and accusative morphology) with a verb is enough context to set up the potential double role. Clearly, context is not to be restricted to larger semantic meaning but would include any well-formed syntactical sequence that allows for more than one reading. After all, on the semantic level, "knowing letters" would not be ambiguous at all, insofar as letters are only possible objects, never subjects, of knowledge.

Appendix 4

1. In fifth-century Greek, ἁπλῶς most commonly modifies verbs of speaking. However, the adverb is not exclusively restricted to such contexts. compare, e.g., Thucydides (7.35): "The Athenians, although none of their ships was sunk outright (τῶν νεῶν κατέδυ οὐδεμία ἁπλῶς), had seven ships put out of action . . ." My claim is not that Plato is unusual in his restricted use of ἁπλῶς, but that Aristotle is unusual in extending the range of ἁπλῶς-modified verbs to include εἶναι, γίγνεσθαι, and so on.

2. Among Plato's uses of ἁπλῶς, he only once approaches an ontological claim. In Book II of the *Republic* he presents an argument to the effect that gods cannot change, and so stories by poets that involve transformations of gods are unsuitable for the education of the guardian class. Plato concludes (381c9), "[a god] remains always simply in his own form" (μένει ἀεὶ ἁπλῶς ἐν τῇ αὑτοῦ μορφῇ).

3. *Republic* 438c7–d1: ἐπιστήμη μὲν αὐτὴ μαθήματος αὐτοῦ ἐπιστήμη ἐστὶν . . . ἐπιστήμη δέ τις καὶ ποιά τις ποιοῦ τινος καὶ τινός.

4. There seems to have been in the Academy of Aristotle's youth a well-known difficulty with positing Thirst Itself as a Form. "Thirst" follows the logic of an incomplete predicate. Without some πρός τι specification, any instance of thirst also can be not an instance of thirst. That is, someone could be thirsty for beer but not thirsty for milk. G. E. L. Owen has argued that this ambiguity with incomplete predicates was one motivation for the positing of Forms of πρός τι predicates (e.g., equal, large, small, beautiful, etc.). Needed was some unambiguous paradigm that possessed the predicate in question καθ' αὑτό. Under this interpretation, Aristotle's remark at *Meta.* A (990b15–17), that some Platonic arguments "produce forms of relatives, of which we say that there is no καθ' αὑτό genus," is directed against Platonists who hold a strict καθ' αὑτό / πρός τι distinction among predicates and who then confuse that distinction by positing a set of καθ' αὑτό Forms of πρός τι predicates. See Owen (1986), 165–79; also see Alexander (1891b), 82.8–83.34.

That Forms of πρός τι predicates like "thirst" result in an incoherent class of καθ' αὑτό relatives may not have been foreseen by Plato when he wrote the *Republic*. However, Plato did have another reason for not regarding Thirst Itself in our text as a Form. The incompleteness of the predicate can be disambiguated by two different specifications. Either one can cite a species of thirst (thirst for milk, thirst for beer, etc.), or one can cite a particular instantiation of thirst (thirst in Socrates, thirst in Callias, etc.). In our text of *Republic* IV, the concern is to specify thirst by species rather than by particular instantiations. Elsewhere, though, when the Forms are explicitly under discussion, the correlate to "f-Itself" is "f-in-us" or "f-in-Simmias" (e.g., *Phaedo* 102–103). This suggests that if αὐτὸ τὸ διψῆν were to be considered a Form, the properly related sensible particular would be τὸ διψῆν in some individual. In fact, not to posit the relationship of Form to particular as Forms ἐν ἡμῖν lands the whole metaphysical doctrine into "the greatest difficulty" of *Parmenides* 133b–134c. That is, the Forms, which are supposed to be the objects of knowledge *par excellence*, are rendered unknowable.

5. *Top.* II, 11, 115b29–35, emphasis added.

6. I take the use of δοκῇ εἶναι in this restatement to be a more cautious variant of ἐστι. To suppose that Aristotle is just here introducing a distinction between what is f and what only appears to be f would be an inexplicable change of issues. Moreover, it would render the conclusion of this bit of argumentation a *non sequitur*.

7. This takeover is not complete. Even in his mature writings, one finds Aristotle using ἁπλῶς as a modifier of speech: e.g., *EN* II, 3, 1104b24–26. Compare this example of ἁπλῶς to others from *EN* in Lewis (1960), 167–70.

8. See Appendix 3.

Bibliography

Ackrill, J. L. 1963. *Aristotle's Categories and de Interpretatione*. Translated with Notes and Glossary. Oxford: Clarendon Press.

Adkins, A. W. H. 1985. *Poetic Craft in the Early Greek Elegists*. Chicago: University of Chicago Press.

Alexander of Aphrodisias. 1891a. *in Aristotelis Topicorum*. (*Commentaria in Aristotelem Graeca*, vol. 2, part 1.) Edited by M. Wallies. Berlin: George Reimer.

———. 1891b. *in Aristotelis Metaphysica*. (*Commentaria in Aristotelem Graeca*, vol. 1.) Edited by M. Hayduck. Berlin: George Reimer.

Ps.-Alexander of Aphrodisias (= Michael Ephesius?). 1898. *in Aristotelis Sophisticos Elenchos commentarium*. (*Commentaria in Aristotelem Graeca*, vol. 2, part 3.) Edited by M. Wallies. Berlin: George Reimer.

Allen, D. J. 1954. "The Problem of Cratylus." *The American Journal of Philology* 75: 271–87.

Allen, R. E. 1960. "Participation and Predication in Plato's Middle Dialogues." *The Philosophical Review* 69: 147–64.

Allen, W. S. 1948. "Ancient Ideas on the Origin and Development of Language." *Transactions of the Philological Society of London* 46: 35–60.

Anton, John P. 1968. "The Aristotelian Doctrine of *Homonyma* in the *Categories* and Its Platonic Antecedents." *Journal of the History of Philosophy* 6: 315–26.

———. 1969. "Ancient Interpretations of Aristotle's Doctrine of *Homonyma*." *Journal of the History of Philosophy* 7: 1–18.

Arens, Hans. 1984. *Aristotle's Theory of Language and Its Tradition*. "Amsterdam Studies in the Theory and History of Linguistic Science," vol. 29. Amsterdam: John Benjamins.

Aristotle. 1886. *Fragmenta*. 3rd ed. Edited by Valentinus Rose. Leipzig: Teubner.

———. 1926. *The "Art" of Rhetoric*. Translated by John Henry Freese. Loeb Classical Library. Cambridge: Harvard University Press.

———. 1949. *Categoriae et Liber de Interpretatione.* Revised text by L. Minio-Paluello. Oxford: Oxford University Press.

———. 1958. *Topica et Sophistici Elenchi.* Revised text by W. D. Ross. Oxford: Oxford University Press.

———. 1959. *Ars Rhetorica.* Revised text by W. D. Ross. Oxford: Oxford University Press.

———. 1960. *Posterior Analytics* and *Topica.* Translated by Hugh Tredennick and E. S. Forster. Loeb Classical Library. Cambridge: Harvard University Press.

———. 1964. *Analytica Priora et Posteriora.* Revised text by W. D. Ross. Oxford: Oxford University Press.

———. 1978. *On Sophistical Refutations.* Translated by E. S. Forster. Loeb Classical Library. Cambridge: Harvard University Press.

———. 1984. *The Complete Works of Aristotle.* Revised Oxford translation. Edited by Jonathan Barnes. Princeton: Princeton University Press.

Artmann, Benno. 1989. "Review of *Das mittelalterliche Zahlenkampfspiel* and *Rhythmomachia.*" *The Mathematical Intelligencer* 11: 77–79.

Balme, D. M. 1972. *Aristotle's de Partibus Animalium I and de Generatione Animalium I.* Oxford: Clarendon Press.

Bar-Hillel, Yehoshua. 1964. "More on the Fallacy of Composition." *Mind* 73: 125–26.

Barnes, Jonathan. 1975a. "Aristotle's Theory of Demonstration." *Articles on Aristotle: Vol. 1.* Edited by Jonathan Barnes, Malcolm Schofield, and Richard Sorabji. London: Duckworth.

———. 1975b. *Aristotle's Posterior Analytics.* Oxford: Clarendon Press.

Basu, Dilip K. 1986. "A Question of Begging." *Informal Logic* 8: 19–26.

Boehner, P. 1958. *Collected Articles on Ockham.* St. Bonaventure, N. Y.: Franciscan Institute.

Bolton, R. 1976. "Essentialism and Semantic Theory in Aristotle." *Philosophical Review* 85: 514–44.

Bonitz, H. 1955. *Index Aristotelicus.* 2d ed. Graz: Akademische Druck—U. Verlagsanstalt.

Bueno, Anibal A. 1988. "Aristotle, the Fallacy of Accident, and the Nature of Predication: A Historical Inquiry." *Journal of the History of Philosophy* 26: 5–24.

Burnyeat, M. F. 1979. "Conflicting Appearances." *Proceedings of the British Academy* 65: 69–111.

———. 1981. "Aristotle on Understanding Knowledge." Pp. 97–139 in *Aristotle on Science: The "Posterior Analytics,"* ed. E. Berti. Padova: Antenore.

Cole, R. 1965. "A Note on Informal Fallacies." *Mind* 74: 432–33.

Colson, F. H. 1919. "The Analogist and Anomalist Controversy." *The Classical Quarterly* 13: 24–36.

Cope, E. M. 1867. *An Introduction to Aristotle's Rhetoric*. London: Macmillan.

———. 1877. *The Rhetoric of Aristotle with a Commentary I-III*. Revised and edited by J. E. Sandys. Cambridge: Cambridge University Press.

Copi, Irving M. 1982. *Introduction to Logic*., 6th ed. New York: Macmillan.

Couloubaritsis, Lambros. 1989. "Dialectic, Rhetoric, and Critique in Aristotle." Pp. 95–110. *From Metaphysics to Rhetoric*, ed. by Michel Meyer. Dordrecht: Kluwer.

De Morgan, Augustus. 1847. *Formal Logic*. London: Taylor and Walton.

Demos, Raphael. 1959. "Partly So and Partly Not So." *Mind* 68: 51–56.

Diodorus Sicilus. 1950. *History*. Translated by C. H. Oldfather. Loeb Classical Library. Cambridge: Harvard University Press.

Diogenes Laertius. 1972. *Lives of Eminent Philosophers*. Translated by R. D. Hicks. Loeb Classical Library. Cambridge: Harvard University Press.

Dorion, Louis-Andre. 1995. *Aristote: Les refutations sophistiques*. Paris: J. Vrin.

Driscoll, John A. 1981. "ΕΙΔΗ in Aristotle's Earlier and Later Theories of Substance." Pp. 129–59. *Studies in Aristotle*, ed. Dominic J. O'Meara. Washington, D. C.: Catholic University of America Press.

Edmonds, John Maxwell. 1961. *The Fragments of Attic Comedy*. 3 vols. Leiden: E. J. Brill.

Evans, J. D. G. 1974. "Aristotle on Relativism." *The Philosophical Quarterly* 24: 193–203.

———. 1975. "The Codification of False Refutations in Aristotle's *De Sophisticis Elenchis*." *Proceedings of the Cambridge Philological Society* 201: 45–52.

———. 1977. *Aristotle's Concept of Dialectic*. Cambridge: Cambridge University Press.

Fair, Frank. 1973. "The Fallacy of Many Questions: Or, How to Stop Beating Your Wife." *Southwestern Journal of Philosophy* 4: 89–92.

Fearnside, W. Ward, and William Holther. 1959. *Fallacy: The Counterfeit of Argument*. Englewood Cliffs, N. J.: Prentice Hall.

Frede, Michael. 1980. "The Original Notion of Cause." Pp. 217–49. *Doubt and Dogmatism: Studies in Hellenistic Epistemology*, ed. M. Schofield, M. Burnyeat, and J. Barnes. Oxford: Clarendon Press.

Galston, Miriam. 1982. "Aristotle's Dialectic, Refutation, and Inquiry." *Dialogue* 21: 79–93.

Geach, P. T. 1972. *Logic Matters*. Oxford: Basil Blackwell.

———. 1976. *Reason and Argument*. Oxford: Basil Blackwell.

Gelber, Hester G. 1984. "I Cannot Tell a Lie: Hugh Lawton's Critique of Ockham on Mental Language." *Franciscan Studies* 44: 141–79.

———. 1987. "The Fallacy of Accident and the 'Dictum de omni': Late Medieval Controversy over a Reciprocal Pair." *Vivarium* 25: 110–145.

Gellius, Aulus. 1927. *Attic Nights*. Translated by John Rolf. Loeb Classical Library. London: Heinemann.

Gershenson, Daniel E., and Daniel A. Greenberg. 1961. "Melissus of Samos in a New Light: Aristotle's *Physics* 186a10–16." *Phronesis* 6: 1–9.

Gillespie, C. M. 1913–1914. "The Logic of Antisthenes." *Archiv fur Geschichte der Philosophie* 26: 479–500; 27: 17–38.

Giuliani, Alessandro. 1972. "The Aristotelian Theory of the Dialectical Definition." *Philosophy and Rhetoric* 5: 129–42.

Graeser, Andreas. 1977. "On Language, Thought, and Reality in Ancient Greek Philosophy." *Dialectic* 31: 359–88.

Grimaldi, William. 1988. *Aristotle, Rhetoric II, A Commentary*. New York: Fordham University Press.

Grube, G. M. A. 1950. "Antisthenes Was No Logician." *Transactions and Proceedings of the American Philological Association* 81: 16–27.

Guthrie, W. K. C. 1969. *A History of Greek Philosophy, vol. 3*. Cambridge: Cambridge University Press.

Guttenplan, Samuel. 1974. *Mind and Language*. Edited by Samuel Guttenplan. Wolfson College Lectures 1974. Oxford: Clarendon Press.

Hamblin, C. L. 1970. *Fallacies*. London: Methuen.

Hamlyn, D. W. 1961. "Aristotle on Predication." *Phronesis* 6: 110–26.

———. 1977. "Focal Meaning." *Proceedings of the Aristotelian Society* 78: 1–18.

———. 1990. "Aristotle on Dialectic." *Philosophy* 65: 465–76.

Hawtrey, R. S. W. 1981. *Commentary on Plato's Euthydemus*. Philadelphia: American Philosophical Society.

Herodotus. 1946. *History*. Translated by A. D. Godley. Loeb Classical Library. Cambridge: Harvard University Press.

Hintikka, Jaakko. 1973. *Time and Necessity: Studies in Aristotle's Theory of Modality*. Oxford: Clarendon.

Irwin, T. H. 1981. "Homonymy in Aristotle." *Review of Metaphysics* 34: 523–44.

———. 1982. "Aristotle's Concept of Signification." Pp. 241–66 in *Language and Logos: Studies in Ancient Greek Philosophy Presented to G. E. L. Owen*, ed. Malcom Schofield and Martha Craven Nussbaum. Cambridge: Cambridge University Press.

Joachim, H. H. 1922. *Aristotle: On Coming-to-Be and Passing-Away*. Revised text with Introduction and Commentary. Oxford: Clarendon Press.

Joseph, H. W. B. 1916. *An Introduction to Logic*. Rev. 2d ed. Oxford: Clarendon Press.

Kahn, Charles H. 1973. "Language and Ontology in the *Cratylus*." Pp. 152–77 in *Exegesis and Argument: Studies in Greek Philosophy Presented to Gregory Vlastos*, ed. Edward N. Lee, Alexander Mourelatos, and Richard Rorty. Assen: Van Gorcum.

Kemp, Alan. 1986. "The *TEKHNE GRAMMATIKE* of Dionysius Thrax." English translation with Introduction and Notes. *Historiographia Linguistica* 13: 343–63.

Kirk, Geoffrey S. 1951. "The Problem of Cratylus." *The American Journal of Philology* 72: 225–53.

Kirwan, Christopher. 1979. "Aristotle and the So-called Fallacy of Equivocation." *Philosophical Quarterly* 29: 35–46.

Kneale, W., and M. Kneale. 1962. *The Development of Logic*. Oxford: Clarendon Press.

Landesman, Charles. 1961. "Does Language Embody a Philosophical Point of View?" *Review of Metaphysics* 14: 617–36.

Lear, Jonathan. 1980. *Aristotle and Logical Theory*. Cambridge: Cambridge University Press.

———. 1988. *Aristotle: The Desire to Understand*. Cambridge: Cambridge University Press.

Levin, Saul. 1983. "The Origin of Grammar in Sophistry." *General Linguistics* 23: 41–47.

Lewis, C. S. 1960. *Studies in Words*. Cambridge: Cambridge University Press.

Liddell, Henry, and Robert Scott. 1940. *A Greek-English Lexicon*. 9th ed. Oxford: Clarendon Press.

Lucas, D. W. 1968. *Aristotle: Poetics*. With Introduction, Commentary, and Appendixes. Oxford: Clarendon Press.

Mackie, J. L. 1974. *The Cement of the Universe: A Study of Causation*. Oxford: Clarendon Press.

McCumber, John. 1982. "Discourse and Psyche in Plato's *Phaedrus*." *Apeiron* 16: 27–37.

McKirahan, Richard D. 1992. *Principles and Proofs*. Princeton: Princeton University Press.

Mill, John Stuart. 1843. *A System of Logic*. London: Longmans Green.

Moravcsik, J. M. 1975. "*AITIA* As Generative Factor in Aristotle's Philosophy." *Dialogue* 14: 622–38.

Owen, G. E. L. 1986. *Logic, Science, and Dialectic*. Ithaca, N. Y.: Cornell University Press.

Patzig, G. 1959. "Aristotle and Syllogisms from False Premisses." *Mind* 67: 186–92.

———. 1968. *Aristotle's Theory of the Syllogism*. Dordrecht: Reidel.

Peter of Spain. 1990. *Language in Dispute*. Translation of *Summulae Logicales* by Francis P. Dinneen. Amsterdam Studies in the Theory and History of Linguistic Science. Amsterdam: John Benjamins.

Pinborg, Jan. 1975. *Current Trends in Linguistics*, vol. 13. "Historiography of Linguistics." Edited by Thomas Sebeok. The Hague: Mouton and Co.

Plato. 1900–1907. *Platonis Opera*. 5 vols. Edited by John Burnett. Oxford: Clarendon Press.

Polybius. 1923. *Histories*. Translated by W. R. Paton. Loeb Classical Library. Cambridge: Harvard University Press.

Poste, Edward. 1866. *Aristotle on Fallacies, The Sophistici Elenchi: With Translation and Notes*. London: Macmillan and Co.

Robins, R. H. 1951. *Ancient and Mediaeval Grammatical Theory in Europe*. London: G. Bell and Sons.

———. 1957. "Dionysius Thrax and the Western Grammatical Tradition." *Transactions of the Philological Society of London* 55: 67–106.

———. 1990. *A Short History of Linguistics*. 3rd ed. New York: Longman.

Robinson, Richard. 1941. "Ambiguity." *Mind* 50: 140–55.

———. 1969a. "The Theory of Names in Plato's *Cratylus*." Pp. 118–38 in *Essays in Greek Philosophy*, Oxford: Clarendon Press.

———. 1969b. "A Criticism of Plato's *Cratylus*." Pp. 100–17 in *Essays in Greek Philosophy*. Oxford: Clarendon Press.

———. 1971. "Begging the Question, 1971." *Analysis* 31: 113–17.

Ross, W. D. 1924. *Aristotle's Metaphysics: Revised Text with Introduction and Commentary*. Oxford: Clarendon Press.

———. 1949. *Aristotle's Prior and Posterior Analytics: Revised Text with Introduction and Commentary*. Oxford: Clarendon Press.

Rowe, W. L. 1962. "The Fallacy of Composition." *Mind* 71: 87–92.

Sanford, David H. 1977. "The Fallacy of Begging the Question: A Reply to Barker." *Dialogue* 16: 485–98.

Schenkeveld, D. M. 1984. "Stoic and Peripatetic Kinds of Speech Act and the Distinction of Grammatical Moods." *Mnemosyne* 37: 291–351.

Scott, Robert. See Liddell, Henry.

Sidgwick, Alfred. 1882. "The Localisation of Fallacy." *Mind* 8: 55–64.

Sim, May, ed. 1999. *From Puzzles to Principles? Essays on Aristotle's Dialectic.* Lanham: Lexington Books.

Simonson, Solomon. 1944. "The Aristotelian Forms of Disputation." *The New Scholasticism* 18: 385–90.

Smith, David E., and Clara C. Eaton. 1911. "Rithmomachia, the Great Medieval Number Game." *The American Mathematical Monthly* 18: 73–80.

Smith, Robin. 1989. *Aristotle: Prior Analytics.* Indianapolis: Hackett.

Smyth, Herbert W. 1920. *Greek Grammar.* Cambridge: Harvard University Press. Revised by Gordon Messing, 1956.

Sommers, Fred. 1969. "Do We Need Identity?" *The Journal of Philosophy* LXVI: 499–504.

Sontag, Frederick. 1954. "The Platonist's Concept of Language." *Journal of Philosophy* 51: 823–30.

Sophonias? (c. 1300). 1884. *Anonymi in Aristotelis Sophisticos Elenchos Paraphrasis.* (*Commentaria in Aristotelem Graeca*, vol. 23, part 4.) Edited by Michael Hayduck. Berlin: George Reimer.

Sprague, Rosamond Kent. 1967. "Logic and Literary Form in Plato." *The Personalist* 48: 560–72.

———. 1977. "Plato's Sophistry." *The Aristotelian Society.* Supplementary Volume 60: 45–61.

Stewart, M. A. 1977. "Plato's Sophistry." *The Aristotelian Society.* Supplementary Volume 60: 21–44.

Stocks, J. L. 1933. "The Composition of Aristotle's Logical Works." *The Classical Quarterly* 27: 115–24.

Thompson, E. Seymer. 1901. *Plato, Meno, with Introduction and Notes.* Cambridge: W. Heffer and Sons.

Trentman, John. 1970. "Ockham on Mental." *Mind* 79: 586–90.

Walton, Douglas. 1981. "The Fallacy of Many Questions." *Logique et Analyse* 95–96: 291–313. Reprinted in Woods and Walton, 1989.

Ware, James H. 1987. "The *Cratylus* and How Words Are Used." Pp. 91–114 in *Contemporary Essays on Greek Ideas: The Kilgore Festschrift*, ed. R. Baird, W. Cooper, E. Duncan, and S. Rosenbaum. Waco: Baylor University Press.

Wedin, Michael. 1973. "A Remark on *Per Se* Accidents and Properties." *Archiv fur Geschichte der Philosophie* 55: 30–35.

———. 1984. "Singular Statements and Essentialism in Aristotle." *Canadian Journal of Philosophy,* Supplementary Volume 10: 67–88.

———. 1988. *Mind and Imagination in Aristotle.* New Haven: Yale University Press.

Whately, Richard. 1854. *Elements of Logic.* Rev. 9th ed. Boston: James Munroe.

William of Sherwood. 1966. *Introduction to Logic.* Translated and edited by Norman Kretzmann. Minneapolis: University of Minnesota Press.

Williams, C. J. F. 1982. *Aristotle's de Generatione et Corruptione.* Translated with Notes. Oxford: Clarendon Press.

Wittgenstein, Ludwig. 1976. *Philosophical Investigations.* Translated by G. E. M. Anscombe. Oxford: Basil Blackwell.

Woods, John, and Douglas Walton. 1977. "Composition and Division." *Studia Logica* 36: 381–406. Reprinted in 1989.

———. 1982. "The Petitio: Aristotle's Five Ways." *The Canadian Journal of Philosophy* 12: 77–100.

———. 1989. *Fallacies, Selected Papers 1972–1982.* Dordrecht: Foris Publications.

Index of Names

Subject Index